The Physiology of Work

The Physiology of Work

Professor Kaare Rodahl
former Director
Institute of Work Physiology, Oslo

Taylor & Francis
London   New York   Philadelphia
1989

UK　　　Taylor & Francis Ltd, 4 John St, London WC1N 2ET

USA　　Taylor and Francis Inc., 1900 Frost Road, Suite 101, Bristol, PA 19007

British Library Cataloguing in Publication Data

Rodahl, Kaare
　　The physiology of work.
　　1. Man. Work. Physiological aspects
　　I. Title
　　612'.042
　　ISBN 0-85066-478-0
　　ISBN 0-85066-483-7

Library of Congress Cataloging-in-Publication Data
is available

Cover design by Barking Dog Art, East Grinstead, UK
Typeset in 11/12 point Bembo by
Mathematical Composition Setters Ltd, Salisbury, UK
Printed in Great Britain by
Taylor & Francis (Printers) Ltd, Basingstoke, Hampshire

Contents

Chapter 1
The biological basis for human work

The human body is made for physical work. This was true in the case of the Stone Age hunter, and it is still the case in the present-day technocrat, for by all indications, the human body has not changed significantly during the last 100,000 years. We are still born with a mass of muscles equivalent to more than half our body weight for the purpose of carrying our body around and for doing muscular work. These muscles are capable of transferring chemical energy in the food which we eat, into mechanical work, with a mechanical efficiency comparable to that of a modern automobile engine. These muscles are not only capable of developing great muscular strength and of carrying loads equivalent to that of our own body weight, they are also capable of generating a wide range of tensions necessary when moving different parts of our body, especially the extremities, and manipulating minute objects, as in the case of electronic technology.

We are, like all higher animals, basically designed for mobility, which is the basic requirement for the performance of any kind of physical work. Consequently, our locomotive apparatus, i.e. the neuromuscular-skeletal system, constitutes the majority of our total body mass.

The muscle, which is the basic instrument of mobility, is a very old tissue. The earliest fossil traces of animal life were burrows which began to appear in rocks almost 700 million years ago (Valentine, 1978). Evidently, these animals could dig in the sea bed, using muscle force. The energy-yielding processes in these muscle cells were the same as those occurring in the first unicellular organisms some 1,500 million years ago, and as those occurring in our muscle cells today (for references see Åstrand and Rodahl, 1986).

Of all the tissues in our body, the muscle is unique in that it can vary its metabolic rate to a greater extent than any other tissue, when changing from rest to work. Skeletal muscles may increase their oxidative processes some 50 times the resting level. While this level of exertion is typical of many types of athletic events, it is hardly ever encountered in ordinary physical work or labour, where the energy metabolism seldom exceeds 25 per cent of the maximal attainable level.

Another feature of the muscle cell is that it, similar to certain other cells, such as nerve cells, is not replaced if destroyed. Contrary to cells such as connective tissue cells, muscle cells do not regenerate. They can, however, increase in thickness by training, and thus develop greater strength.

The supply of blood, and thereby the supply of nutrients and oxygen, to the working muscle cell is determined by the amount of work the muscle is doing. When resting, the smooth muscle fibers constricting the blood capillaries supplying the skeletal muscle tissue, are contracted, keeping the capillary lumen nearly closed, so that the muscle cells only receive a minimum of blood, sufficient to stay alive. When working, however, the active muscle cells produce metabolites which relax the smooth capillary muscles, and allow more blood to pass into the muscle tissue; and more so the heavier the muscle is working. This is an example of one of nature's decentralized, democratic regulatory mechanisms based on the principle that the muscle cells which are most active receive the greatest share of the available blood supply. Thus, in the case of the muscle, it pays to work: the more they do, the more they get.

Some 10–20 million years ago our forerunners, the hominids, developed as a separate branch from the family tree of primates. When they abandoned their life in the trees and started to forage and hunt on the ground, they initiated a bipedal adaptation to terrestrial life (Valentine, 1978). Their pelvis permitted an upright posture with a bipedal gait, which freed their arms and hands for carrying their necessities with them as they walked. Tool-making was probably established more than three million years ago, before any marked brain expansion had occurred (Lewin, 1981). By the time of the Neanderthal man appearing more than 100,000 years ago, the same postural abilities, manual dexterity, and range and character of movement typical for modern man, had been developed. Up to that time the increase in brain size seems to be correlated with the increasing complexity of stone tools (Washburn, 1978).

Just as the proportions of the human hand, with its opposable thumb, is ideally suited for manual work and the use of tools, so does the anatomy of the human brain reflect a successful adaptation for manual and intellectual skills. The superiority of the human brain is clearly evidenced by the proportion of the total energy production devoted to the central nervous system. While adults of most vertebrate species devote less than 8 per cent of their basal metabolism to the central nervous system, our central nervous system consumes 20 per cent of our basal metabolism (Mink *et al.*, 1981). The superior human brain capacity not only makes us ideally suited for intellectual work, but also highly suited for skilled manual and physical performance. Consequently, an exceptionally large area of our motor cortex is devoted to the motor control of our hands and fingers. It is a remarkable fact that from all indications this is not a recent evolutionary feature. Most likely, a human being living more than 50,000 years ago had the same potential for physical and intellectual performance, including constructing and operating a computer, as anyone of us living today.

It thus appears that ancient man, as well as modern man, was basically made to employ his body physically in his struggle for survival. Man, like all higher animals, reacts to almost any kind of a threat by mobilizing his resources for a physical effort, i.e. by the so-called stress reaction. The sight or awareness of any kind of threat automatically evokes nervous and hormonal reactions, mobilizing the bodily resources for a physical response to the threat (Cannon, 1929). The pulmonary ventilation increases as does the heart rate and cardiac output, increasing the oxygen uptake and the oxygen delivery to the muscles, in readiness for work. Much of the circulating blood volume is shifted from the guts and skin to the muscles. Due to the effect of some of the stress hormones liberated as a consequence of the stress mechanism, stored energy is released in the form of glucose and free fatty acids, now being made available as substrates for the working muscles. At the same time the excreted stress hormone Cortisol exerts a profound effect on a number of cellular functions, including cellular membrane permeability and the breakdown of proteins to carbohydrates to be used as metabolic fuel for the working muscles (for references, see Malm, 1969; Francis, 1979; Åstrand and Rodahl, 1986).

As already pointed out by Cannon (Herd, 1972), the purpose of all these changes, brought about by the stress reaction serving as a trigger mechanism, is primarily to prepare the organism for a physical effort to cope with the threatening situation – as was the case when a Stone Age hunter suddenly encountered a threatening bear in the wilderness. As soon as the threatened individual engaged himself in the physical activity which the situation called for, be it a fight with the bear or a run for safety, the inherent physiological regulatory mechanisms of the stressed body took over, adjusting the regulatory organ systems to meet the need of the body in terms of physical performance, i.e. pulmonary ventilation, cardiac output, blood volume distribution, fuel mobilization, temperature regulation, etc. Similarly, as soon as the effort was over, the same regulatory mechanisms brought the excited organism back to a state of rest and recuperation (Åstrand and Rodahl, 1986).

Our problem is that any stressful situation in our civilized existence may trigger the same stress reaction as in our primitive ancestor confronted with a wild beast, but without being able to achieve a physical outlet for our stress response. This, in the long run, may cause ill effects, including gastro-intestinal upset and elevated blood pressure (see Chapter 6).

If it may be assumed that the remaining primitive groups of people – such as the Bushmen in Africa, the Aborigines in Australia and the Eskimos in the Arctic – reflect the behaviour of our original ancestors, a great deal can be learned about our functional biology from watching them and the way they engage their endowed body potential in their struggle for survival. One feature which they all seem to have in common is that they move about at a fairly moderate pace, when they hunt or gather food. With great endurance they may stand or sit patiently for hours waiting for the game to appear, but seldom run, at any rate not for very long distances. This is in direct contrast to our own emphasis on the practice of athletic activities such as prolonged jogging or even marathon running. It is true that the Eskimos, and especially the Eskimo children and youngsters, may engage in many kinds of competitive sports, but in most cases they involve brief spurts of dynamic activity, interspaced with periods of rest and relaxation. Perhaps the most vigorous and protracted athletic performance of the Eskimo is their drum dance, a highly dynamic and varied physical activity.

Basically, most of our inherent physiological functions involve rhythmic, dynamic activity, interspaced with inactivity or relaxation, i.e. rhythmic contraction and relaxation of the respiratory muscles as we breath, the cardiac muscle pumping the blood, the skeletal muscle as we walk, the smooth muscles of the gut causing peristaltic movement of its content, and the repeated electrical discharges in the nervous system causing the initiation and propagation of the nervous impulses.

It is thus evident that our body machinery is ideally suited for dynamic work in which exertion and relaxation interchange more or less rhythmically, with a varied use of the different parts of the body, both physical and mental. In the final analysis this may even apply to the interplay between physical and mental engagement. While this type of varied body use was commonplace among our primitive ancestors, it is certainly not so in the case of more sedentary factory workers, such as those engaged in electronic assembly work. The food forager was free to move about, using his eyes to view the horizon or to see where he was going, moving his head and his extremities in a varied way, changing the load on his muscles and constantly alternating the use of his body. In this manner he was able to rest parts of his body while engaging other parts as he proceeded on his way. In contrast to this, our modern electronic assembly line operator, often a woman, usually has to focus her attention on the minute microchips or other parts in front of her. In order to keep the tiny part in sight, she has to keep her eye muscles constantly tensed, and her head fixed, necessitating a constant tensing of the muscles of her neck. In order to move the delicate instruments manipulating the minute parts that she is working on, she has to keep the muscles of her shoulder, arms and hands fixed and tensed all the time, perhaps for hours, day after day, month after month. This sooner or later may lead to pain which in some cases may be the cause of disabling complaints and prolonged sick leave, and varying degrees of disability.

A study of an electronic assembly plant in Norway (Westgaard *et al.*, 1984), revealed that 60 per cent of the healthy young girls starting to work in the plant developed muscular-skeletal complaints causing sick absenteeism for three weeks or more at the end of two and a half years. Even when the work place and the work procedures were improved to meet all ergonomic requirements, the symptoms still developed, the only difference being that they occurred a little later.

The conclusion which may be drawn from this brief review of the biological basis for human performance is that we are ideally suited for varied dynamic work, and that many of the problems of the present-day worker are the consequences of too much static work and stress, with which we are unable to cope.

Chapter 2
Work physiology

Originally, the science of physiology was largely concerned with the function of the healthy organism at rest. Only in recent decades has the student of physiology and medicine been concerned, to any extent, with the function of the human body under the stress of work.

Yet, there were exceptions. Probably one of the very first to emphasize the importance of the study of work, as a means to attaining safe and proper working conditions, was the Polish educator Wojciech Jastrzebowski, who died in 1882 (Polish Ergonomics Society, 1979). During the early part of this century a few of the physiologists in Europe, especially in Germany and England, were actively engaged in the study of several fundamental aspects of the physiology of work and exercise. The first Institute of Work Physiology was founded by Max Rubner in Berlin in 1913, 'Kaiser-Wilhelm-Institut für Arbeits-physiologie' (Rutenfranz et al., 1981), later (1929) known as the Max-Planck-Institute of Work Physiology in Dortmund, under the leadership of Gunther Lehmann, the author of one of the first major textbooks of Applied Work Physiology (Lehmann, 1953). Significant contributions to this field were also made by A. Krogh and his co-workers in Copenhagen, eventually extending to the other Scandinavian countries. In Great Britain, A.V. Hill made a series of fundamental contributions to the basic understanding of the functioning of the human 'living machinery' from 1925 to the beginning of World War II (Jokl, 1980). In the USA a group headed by D.B. Dill formed the Harvard Fatigue Laboratory which made significant contributions to the field of work physiology until it was dissolved during the Second World War.

During the war, applied human factors scientists became involved in what was called human engineering. They were trying to adjust physical facilities, such as the aircraft cockpit to fit the human body which was to occupy it, and arranging the instrument panel in a manner most suitable for the operator in terms of convenience and efficiency. Many of them were psychologists, but the working team might also include physiologists, anthropologists and engineers.

This type of activity continued after the war both in the USA and

Europe under the name of Human Factors Research, mostly in military and aerospace laboratories. The problem areas included simulator operations, cockpit design, communication systems, stress response, etc. In the early 1950s this became known as ergonomics, a term which by some workers in the field is taken to mean more or less anything that has to do with work, including the entire subject of work physiology, work psychology, as well as human factors and human engineering.

During the Second World War some of the extreme environments facing the fighting soldier (such as the heat of the desert), or the sailor (such as the operation of manned submarines) captured the interest of many capable physiologists (see Adolph, *Physiology of Man in the Desert*, 1947, as an example). This type of occupational and environmental physiology under rather extreme conditions (Arctic warfare, altitude and underwater operations) attracted considerable support and recruitment also during the first few decades after the war. This was especially the case during the Cold War prior to the development of intercontinental missiles when the shortest distance by air between the two opposing superpowers was across the Arctic Ocean. At that time, cold weather operations and Arctic survival were matters of prime importance. This brought out the need for a better insight into the factors affecting human cold-weather operations. Consequently, funds in support of these aspects of work and environmental physiology became abundantly available. As a result of this, groups of very capable work physiologists were formed in a number of well-equipped laboratories in several countries around the world. Eventually, these groups contributed significantly to the understanding of such fundamental problems as temperature regulation, the physiological reaction to cold exposure, etc. When the intercontinental ballistic missiles more or less replaced the need for large numbers of Arctic ground and air forces in constant combat readiness, the need for research on temperature regulation and related fields was no longer obvious, and much of the support for this aspect of work physiology was discontinued. As a consequence, the research effort in these areas has remained at a fairly low level world wide, and the progress made since the active period during the 1950s is comparatively modest.

In its early stages of development, work physiology included both the physiology of physical work or labour, and the physiology of athletic performance. It was all considered work in terms of the basic definition of the term: application of force through a distance. In those days when heavy physical work, taxing a sizeable part of the individual's maximal work capacity, was still commonplace, this concept appeared logical. Consequently, a great deal of emphasis was placed on the energy cost of physical work, and the corresponding requirement for dietary energy intake, rest pauses, endurance, movement (kinesiology), the assessment of maximal physical work capacity, and the fraction of this capacity which was being taxed in the performance of the work operation in question (for references see Åstrand and Rodahl, 1986). Since the same organ functions were involved, both in physical labour and athletic exercise, it was natural to consider both disciplines as one.

This may no longer be equally true, since very heavy physical work no longer is generally encountered in manual labour. In fact, work loads in excess of 30–40 per cent of the individual's maximal work capacity are very rare. Furthermore, the metabolic events, and the organic stress at the cellular level, are entirely different in the two situations: ordinary manual labour, and competitive athletic exercise. This was already pointed out by Courtice and Douglas in 1936, when they wrote:

'In everyday life the muscular work which we are called upon to perform is done at a far lower rate than that which is done by the athlete who is pushing himself almost to the limit of his power. It by no means follows that the facts observed during severe exercise, and the deductions made from these, can be applied without alteration when we wish to explain the effects of exercise of far less severity.'

It is therefore perhaps not surprising that those studying the physiology of athletes in recent years have insisted on distinguishing between the physiology of work in terms of physical labour, and the physiology of exercise in terms of athletic performance. Of these two research areas, the physiology of exercise has attracted greater interest in terms of recruitment of research workers than has the physiology of work.

Whereas work physiology was originally, to a large extent, concerned with heavy physical work and labour, this is, as already pointed out, no longer the case. Much of the heavy physical labour is already eliminated and much of what is left can quite easily be taken over by mechanical aids.

In this process of abolishing heavy, dynamic physical work, we have gradually replaced it with prolonged static work, using relatively small muscle groups, often in a fixed position, at an intense working rate, involving considerable mental tension and requiring a high degree of accuracy, a stress situation the human organism is not ideally equipped to endure or to cope with.

During recent decades ergonomists have made admirable efforts to improve the working conditions in keeping with the limitations of the human capabilities. They have improved the arrangement of the actual work place in terms of working posture, reach, table heights, and have adjusted the tools to fit the worker who is using them. Yet the net result is an increasing number of workers suffering varying degrees of muscular-skeletal disorders from painful, transient muscle tension, to chronic disabling fibromyalgia. This is the result of exposing the working individual to a type of static, monotonous, prolonged, and repetitive work, at a high level of maintained concentration and neuromuscular tension, for which the human mind and body is not suited. Consequently, there is a limit to how much the ergonomists may achieve in terms of preventive measures. They may, in fact, in some cases, already have reached the limit, in spite of having developed an awareness on the part of the working individual as to how to sit and how to use the tools, etc., for the symptoms still appear in an ever-increasing number. Consequently, instead of preventing these ill effects, the ergonomist is now, to an increasing degree, involved in the treatment of the symptoms of a faulty use of the human body as a working machinery.

Now the major problem, and indeed the most costly one in terms of sick absenteeism, is the consequence of too much prolonged static work stress, which in a large number of cases inevitably will lead to clinical symptoms and complaints, in some cases associated with objective clinical signs, of a disease process. Thus, the problem may then no longer be a question of physiology, but one of pathophysiology or pathology.

This means that the subject of work physiology is moving closer towards the area of industrial medicine and occupational health, a field to which the work physiologists may make considerable contributions by applying their physiological knowledge and their physiological methods and techniques. It is of interest to note that such a merger of work physiology and occupational health is in keeping with the original concept of Max Rubner, expressed in his proposal for the establishment of a Research Institute for Work Physiology in 1911 (Thomas, 1932), in which he suggested that the activity of the institute should include human physical and mental endurance in different occupations, the effect of age, sex, race, individual differences in diet, as well as the effects of the working environment.

Ideally, the work physiologist would like to study physiological functions of the organism subjected to the stress of work in real life, rather than under simulated conditions in the laboratory. This requires portable measuring devices, capable of sensing and logging the parameters in question, without interfering with the individual's freedom of movement or work performance. In most cases this means that the devices have to be carried on the person. The lack of such transportable field equipment which could be carried on the person limited to a large extent the type of field studies which could be made during the early years of work physiology. Devices of this kind were initially limited to a few items such as the Douglas bag, developed by the British physiologist C.G. Douglas (Courtice and Douglas, 1936), and the German *Gasuhr* (see Rutenfranz *et al.*, 1981) which could be used to record pulmonary ventilation and respiratory gas exchange, and thereby energy metabolism. In some instances small vans or buses were equipped with power generators, etc. and used as mobile field laboratories; these buses were taken to different work places for specific studies concerning such aspects as temperature regulation and body heat balance, energy metabolism and metabolic cost of different types of work.

As a result of the space program, sophisticated miniature electronic equipment appeared in the late 1950s, which could be mounted on the person and used to record and transmit such parameters as the electrocardiogram, which was initially based on radio transmittance. This made it possible to record and log physiological parameters such as heart rate, pulmonary ventilation, oxygen uptake, body temperature, etc., on individuals while exposed to the stress of work or the working environment, in the space craft, in the factory or plant, on the mountain peak, or on the athletic field. This was followed by a subsequent generation of loggers based on portable, miniature electro-magnetic tape recorders, and now finally a generation of miniature battery-powered electronic loggers with as many as a dozen inputs where the data can be stored in digital form. These loggers are in turn integrated with portable battery-powered suitcase-size personal computers and print-out systems, enabling an immediate statistical analysis, display or print-out of the results on the spot. In this way it is possible to record the subject's physiological reactions to a variety of stresses or environmental parameters simultaneously, while performing regular work. When this is combined with detailed activity logs, it is possible to observe what the subject is reacting to, how he is reacting, and what consequences it may have.

With the availability of these sophisticated, yet simple to operate instruments, it is now possible to transfer the work physiology laboratory from the academic campus out into the factories or other work places, where basic parameters can be measured and recorded with the same high degree of accuracy as in the basic laboratory. In this manner, it is now possible to record simultaneously environmental parameters such as air temperature, air motion, humidity and radiation, in addition to the concentration of gases such as O_2, CO_2, CO, SO_2, etc., in the air, as well as physiological parameters such as heart rate, muscle tension, deep and superficial body temperatures, etc., without interfering with the subject's normal activities. Thus, it is now possible to obtain a comprehensive, combined picture of the stresses, both external and internal, to which the subject is exposed and how he is reacting to them physiologically.

Obviously, problems related to a specific work stress or work situation are best studied where the work is being performed, i.e. at the actual work place, since it is not always possible to duplicate completely

the work situation or the working environment in a simulated laboratory experiment. With the sophisticated tools now available, this can be
done with the same degree of accuracy and sophistication as in the
laboratory, providing the worker is willing to serve as a subject for the
study, and providing the research worker is willing and able to put up
with the inconvenience of field work. An added advantage is that the
research worker in the course of his work will become thoroughly
familiar with the practical aspects of the problems he is studying as well
as getting some insight into the problems of industry in general. This
may enhance his ability to use his physiological knowledge in general to
solve other practical problems in this kind of industry.

In the past, industry has, on the whole, often been reluctant to expose
their premises to extensive systematic and penetrating surveys of
specific aspects of their working environment and work stresses. In the
cases where this has been done it has, as a rule, been a question of
external experts coming into the plant to carry out certain general
surveys of the work place or the working environment or specific
aspects of it. More often than not these external teams of experts, upon
the completion of the study, have withdrawn with their data, perhaps to
prepare a report to be submitted several months later. This report might
be read by a small number of individuals, often without any consequence or without resulting in any significant changes. Instead, it
would perhaps have been better if these experts had attempted to teach
the industrial supervisors to carry out and interpret certain basic
measurements in their own plant, and to use this information in their
daily decision-making, in order to improve the working environment
and reduce the stress of work for their employees.

Continuous and simultaneous recording of several environmental
stress factors in many types of industrial working environments is not
only desirable to fulfill the requirements of existing official rules and
regulations. It is equally important for the plant's technical managers to
have at their disposal a current up-to-date display of important basic
parameters concerning the working environment as a basis for their
management of the plant. If this logging of current environmental
parameters – such as heat, humidity, air movement, radiation, noise,
dust, vibration, and air pollution – is expanded to include the objective
physiological reactions of the individuals who are exposed to these
environmental stresses as well, both the management and the worker's
representatives will have an objective, precise and factual basis for their
evaluation of the existing conditions, and an objective and impartial
basis for their mutual judgment of the effects of the changes and
improvements which they are making for better health and well being
of the workers, and better productivity for the benefit of the company.

So far it has been technically possible to record and log a number of
parameters in the physical environment as a matter of routine. In

addition, one has been able to record routinely the worker's response to the environment with the aid of a separate miniature portable logging system such as the Vitalog. This has proved to be of interest, both to management and labour. It is now possible, with the aid of the new Squirrel 1200 generation loggers (Grant, Cambridge, England), to combine both the logging of the environmental parameters, and the worker's physiological reactions to them, by one and the same logger, carried on the worker or operator. Such a system will enable us currently to print-out and display on a screen all values of interest, and where these values stand in relation to existing regulations, established norms, or aims set by the company policy.

Chapter 3
Problem areas in the field of work physiology

The primary objective of the work physiologist is to contribute to the understanding of the many factors which may affect an individual's ability to work, i.e. to transform chemical energy into muscular contractions, which may be used to perform physical work. Since some degree of muscular activity is required in all kinds of work, even in most intellectual occupations, and in all expressions of life, work physiology is of interest not only to those concerned with manual labour. Furthermore, some of the factors affecting the energy-yielding processes are not merely physiological in nature, but include a variety of both psychological and clinical aspects, including the worker's state of health and motivation, in addition to the nature of the work itself and the environment in which the work is being done. In certain types of industry, toxic agents in the air at the place of work are not only of interest to the toxicologist, but to the physiologist as well, since the uptake, and therefore the noxious effect of the toxic agents, is dependent upon pulmonary ventilation and other physiological parameters. This is yet another reason for a closer collaboration between the work physiologist and other occupational health specialists.

In the following, we shall attempt to present a brief review of some of the more important factors affecting our ability to perform physical work, based on Figure 3.1.

Service functions

As already pointed out, the ability to perform physical work basically depends on the ability of the muscle cell to transform chemically-bound energy in the food which we ingest, into mechanical energy. This in turn depends on the capacity of the service functions that deliver fuel and oxygen to the working muscle fiber. This involves intake, storage and mobilization of nutrients (fuel), and the uptake of oxygen through

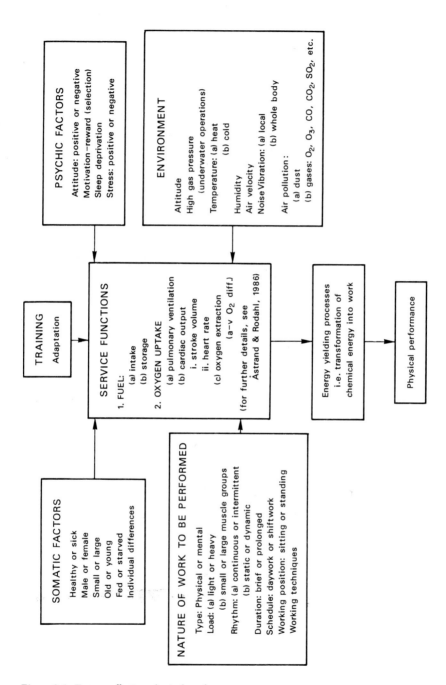

*Figure 3.1. Factors affecting physical performance
(modified from Åstrand and Rodahl, 1986).*

the lungs and the delivery of this oxygen by the cardiovascular system to the muscles for the oxidation of the fuel for the purpose of providing energy. (For details see Åstrand and Rodahl, 1986.)

Training and adaptation

Some of these service functions are subject to improvement by physical training. An important part of this training effect is an increase in the stroke volume of the heart, which causes an increase in the cardiac output, augmenting the oxygen transport from the lungs to the muscles (Saltin *et al.*, 1968; Figure 3.2). A study of assembly-line operators showed that the physically fit workers were less fatigued at the end of the work shift than were those who were unfit (Hettinger and Rodahl, 1960a).

A certain amount of adaptation of the energy-yielding processes, and some of the service functions supporting them, may take place under certain adverse environmental conditions. Thus, exposure to hot

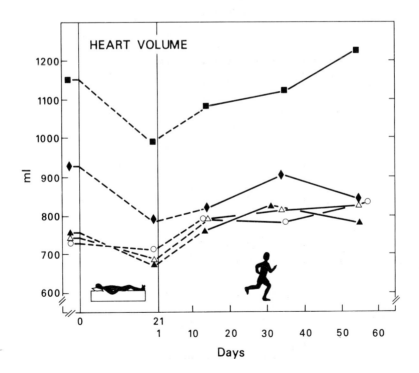

Figure 3.2. Changes in heart volume with bed rest, followed by physical training (from Saltin et al. 1968). Different symbols correspond to individual subjects.

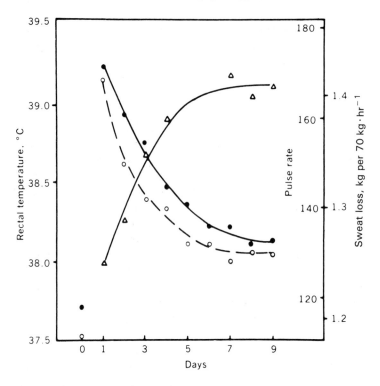

Figure 3.3. Mean rectal temperature (●), heart rate (○), and sweat loss (△) in a group of men during a 9-day acclimatization to heat (modified from Lind and Bass, 1963).

environments may within a week or so cause a beneficial acclimatization to heat, causing better heat tolerance due to more efficient sweating, lowered skin and body temperature, and a reduced heart rate (Lind and Bass, 1963; Figure 3.3). Prolonged exposure to reduced barometric pressure at high altitude leads to compensatory acclimatization changes involving increased pulmonary ventilation, increased haemoglobin concentration in the blood, and increased capillarization, myoglobin content and modified enzyme activity in certain tissues. (For references see Åstrand and Rodahl, 1986.)

Somatic factors

State of health

Physical performance may indeed be affected by the individual's state of health. This relationship, however, is by no means absolute or clear cut.

Thus, remarkable accomplishments have been achieved by highly motivated but handicapped or disabled individuals, who were far from healthy. On the other hand, apparently healthy individuals may feel unable to contribute their share of work, due to a variety of subjective complaints.

Apart from cases of chronic disease, most individual's state of health may be relative. It is not always a matter of being absolutely healthy or completely sick, but of being somewhere in between, or in the process of becoming less well or less sick. Certainly, the state of this process will affect the individual's performance capacity. The point at which the individual feels sufficiently sick to stay away from work, may depend on the type of work to be performed, but may at any rate vary greatly from one person to the next. Yet there are a large number of individuals who suffer from chronic diseases such as cardiovascular disease, pulmonary disorders, gastrointestinal ailments or skeletal-muscular complaints, which may be kept sufficiently under control to enable the victims to do their work on a regular basis. However, some of these ailments may prevent the individual from doing certain jobs or from working under certain environmental conditions, such as excessive heat or abnormal atmospheric pressures.

In the individual assessment of fitness for work in such cases, the clinical work physiologist may be of help in performing specific laboratory tests such as work electrocardiography, or similar stress tests, as well as ambulatory recordings of ECG, heart rate, blood pressure and body temperature on the individual during the performance of his work at the actual work place.

Sex

Basically speaking, there are few jobs which cannot be equally well performed by females as males, except in cases where heavy lifting or very heavy physical work loads are involved. As a matter of fact, experience has shown that women may, in many cases, be superior to men, especially in jobs requiring accuracy, patience, alertness and perseverance. While visual acuity may be slightly better in men, women's colour vision is superior to that of men. Women are better able to identify substances by smell than are men; their hearing is better, especially in the case of high frequency sounds, and they maintain a better hearing as they become older than do men. The women's skin is more sensitive than is the skin of men. (For references see Velle, 1987.) On the basis of the available data there is no difference of practical importance between men and women in their tolerance to hot or cold environments (Nielsen, 1980, Åstrand and Rodahl, 1986, p. 617). The maximal aerobic power, i.e. the maximal oxygen uptake as an expression of an individual's physical work capacity, is not significantly

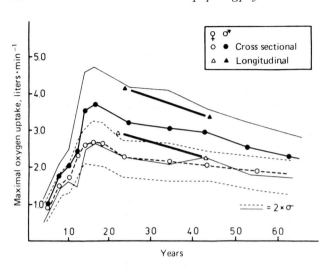

Figure 3.4. Mean values for maximal oxygen uptake (maximal aerobic power) in male and female subjects 4 to 65 years of age (from I. Åstrand et al., 1973a).

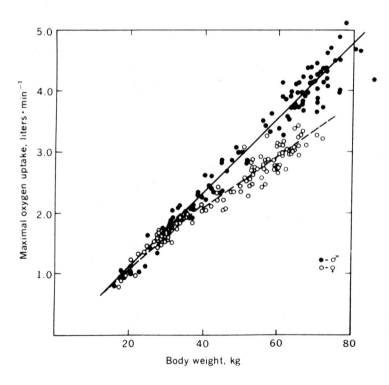

Figure 3.5. Maximal oxygen uptake in 227 male and female subjects, 4 to 33 years of age, in relation to body weight (modified from P.-O. Åstrand, 1952).

different in boys and girls up to the age of puberty (Figure 3.4). Thereafter, the woman's power is, on the average, 65 to 75 per cent of that of the men (I. Åstrand *et al.*, 1973). This means that women exposed to the same physical work load as men may have to exert themselves some 25–30 per cent more than men for the same salary. In the case of light to moderate physical work, however, this may not be of any practical importance.

Women have approximately the same maximal oxygen uptake per kilogram of fat-free body mass as men (Figure 3.5). The lower maximal oxygen uptake for females after the age of about 14 years is largely explained by their higher content of adipose tissue. The lower concentration of hemoglobin in the women's blood also contributes to the observed difference in maximal aerobic power between the sexes (Åstrand and Rodahl, 1986).

Changes in maximal muscle strength with age in women and men are shown in Figure 3.6. The increase in muscle strength in children is definitely a consequence of growth. In children up to the age of ten to twelve years, there is no significant difference in strength between boys and girls. After this age the boys continue to become stronger, while the girls gradually level off. The difference is mainly caused by differences in muscle mass (25–35 per cent of the body weight in women, as against 40–45 per cent in men). Thus, the maximal strength per unit of cross-sectional area of the skeletal muscle is about the same in men and women of all ages. In absolute figures, however, the women's maximal strength of the leg muscles is roughly 65–75 per cent of that of the men. In the case of the trunk muscles the figure is slightly

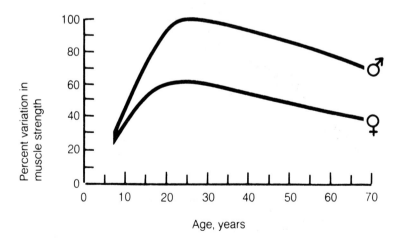

Figure 3.6. Changes in maximal isometric strength with age in women and men (by courtesy of E. Asmussen).

lower (60–70), while in elbow flexion and extension the strength of women is only some 50 per cent of that of males. (For references see Maughan, 1984, Åstrand and Rodahl, 1986.)

Body size

While aerobic power and muscle strength are related to body size and muscle mass, body size as such is rarely a limiting factor in industrial performance. In terms of the energy cost of transporting the body by locomotion, the most efficient vehicle known is a human being on a bicycle, requiring only one-fifth of the energy used when the same human being is walking (Diamond, 1983). Yet, the natural evolution never created the wheel.

Age

Maximal oxygen uptake, heart rate, stroke volume, pulmonary ventilation, and muscle strength decrease significantly with old age (Figure 3.7). In both sexes, the maximal aerobic power reaches a peak at the age of 18–20 years (Figure 3.4), followed by a gradual decline (I. Åstrand *et al.*, 1973). At the age of 65, the mean value is about 70 per cent of what it is for a 25-year old individual. The value for a 65-year old man is, on the average, roughly the same as that of a 25-year old woman.

As is evident from Figure 3.6, a peak is usually reached in maximal muscle strength at the age of about 20 for men, and a few years earlier for women. The strength of a 65-year old individual is, on the average, 75–80 per cent of that attained at the age of 20–30 years. The rate of decline in muscle strength with age is in both sexes greater in the leg and trunk muscles than in the strength of the arm muscles (Grimby and Saltin, 1983). The decline in muscle strength with age is due to a decline in muscle mass, due to a reduction in the number of motoneurons. A decline in muscle strength of about 30 per cent from the age of 30 to the age of 75 coincides with a 30 per cent decline in the urinary elimination of creatinine, suggesting a muscular atrophy of this magnitude (Figure 3.8).

Nutritional state

In carefully controlled laboratory studies varied nutritional states in terms of caloric and protein intakes over a 10-day period had no appreciable effect on physical performance at room temperature. However, a combination of nutritional deficiency and cold stress had deteriorating effects on performance (Rodahl *et al.*, 1962). Similarly,

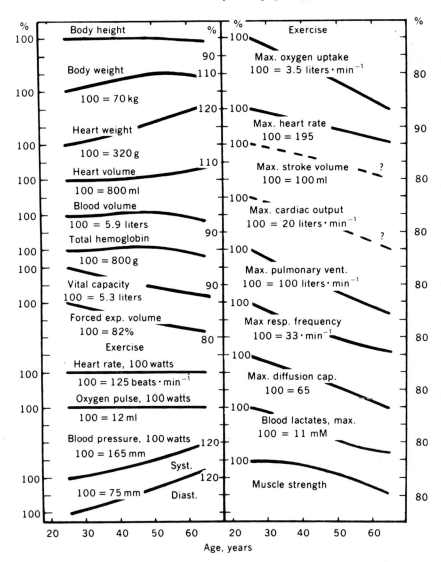

Figure 3.7. Variation in some static and dynamic functions with age (from Åstrand and Rodahl, 1986).

Crowdy (1970) observed no apparent negative effect on performance in troops living on restricted rations for limited periods of time. In a study of military cadets exposed to considerable work stress during a 5-day period on reduced dietary intakes combined with sleep deprivation, it was shown that dietary supplements did not affect performance significantly, while sleep did (Opstad, 1987).

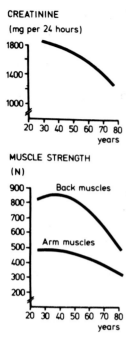

Figure 3.8. Variation in 24-hour urinary excretion of creatinine as an index of the total muscle mass, and muscle strength (from Åstrand and Rodahl, 1986).

Individual differences

Individual differences in special skills or functional capabilities, whether genetically endowed or not, may certainly play a role in many types of occupations where such special skills and capabilities are essential (aircraft pilots etc.). This emphasizes the importance of proper selection procedures.

The magnitude of individual differences in performance capacity may perhaps best be illustrated by considering the wide scatter of such a basic function as the maximal oxygen uptake in a group of individuals of the same sex and of the same age. It may vary from one extreme to the other by a factor of 2 or 3 (Åstrand and Rodahl, 1986).

Psychic factors

Attitude

Obviously, a person's attitude to work in general, and to his job in particular, affects performance. Traditionally, certain jobs may have been looked upon as being typically suited for men, others for women. While this may have changed with changing trends in the industrialized parts of the world, these attitudes may still persist in more primitive societies. At any rate, the individual's willingness to devote himself to his work and to make it a matter of pride to do a good job, varies greatly from one individual to another, and perhaps even from one society to the next.

Motivation

Motivation plays a major role in all kinds of human pursuits. Biologically speaking, the basic concept of reward and punishment still plays a basic role in determining our attitude towards work, and our willingness to employ all our resources in the pursuit of our goals. This aspect of our state of mind is closely associated with an awareness of a personal role in the working team or in the job. A strong sense of responsibility may cause an individual to feel that he would be missed, or that he would let his co-workers down, if he failed to show up for work. In many cases, a system of piece work as opposed to fixed salary, may be an incentive to work harder. In most individuals, a role of responsibility may in itself be a motivating factor. In field trials, involving prolonged strenuous physical activity combined with reduced food intake and sleep deprivation, we observed that giving the subjects responsibility or a leading role caused them to improve their mental performance for brief periods.

Sleep deprivation

Sleep deprivation, when severe, i.e. extending over more than two-to-three days, may cause marked deterioration of mental performance. This includes information-processing, perceptual-motor tasks, word recall, mathematical problem solving, visual vigilance tasks, auditory tasks and memory tasks. The effect is particularly noticeable in the case of sustained or continuous mental tasks (Asken and Raham, 1983).

Emotional effects are even more conspicuous after several days of sleep deprivation, including depression, hostility, anxiety, tension, confusion and even hallucinations. In a study of sleep-deprived military cadets it was found that three hours of sleep caused a marked improvement in performance (Opstad, 1987).

Stress

As will be discussed in more detail in Chapter 6, stress with which one is able to cope, acts as a stimulation, arousing one's mental alertness and intellectual activity, up to a point. Stress with which one is unable to cope, on the other hand, may affect the individual subconsciously, making him preoccupied with the problems involved. This may block his ability to concentrate on the task confronting him, and to cope with the situation which may develop. Examples of this have been seen in wartime fighter pilots during combat, commercial pilots and in people involved in traffic accidents.

Nature of work to be performed

Type of work (physical or mental)

The human being is unique among mammals in that he utilizes an exceptionally high proportion of his overall energy output in his central nervous system (Mink *et al.*, 1981). The energy metabolism of the central nervous system is extremely high, even at rest. Most studies in the past have failed to detect any significant increase in the overall energy metabolism of the body during mental activity such as thinking or problem solving, unless it is associated with some form of physical activity involving muscle contractions. Recent measurements have shown, however, that when subjects are engaged in pure mental activity, regional cerebral oxidative metabolism, in several areas of the brain, increases, and so does regional cerebral blood flow. This also includes the cerebellum. In fact, different types of mental activity cause such increases in blood flow and oxidative metabolism in different parts of the brain (Roland, 1987).

Physical work

Physical work performed by the human body is accomplished by muscular contractions, supported by oxidative metabolic processes in the muscle cells, in which fuel (essentially carbohydrate and fat) is oxidized in the presence of oxygen. For each liter of oxygen utilized, some 5 kcal (about 20 J) of energy is liberated. Oxygen uptake, therefore, is an expression of the rate of energy output or rate of work (Figure 3.9). The mechanical efficiency of the human body as a work-producing machine is roughly 20–25 per cent, or similar to that of an automobile engine.

Since all life functions generally consist of rhythmic, dynamic muscular work in which work and rest, muscle contraction and relaxation, are interspersed at more or less regular, fairly short intervals, the ideal way to perform physical work is to perform it dynamically, with brief work periods interrupted by brief pauses. This routine will provide some rest during the actual work period, so the worker may

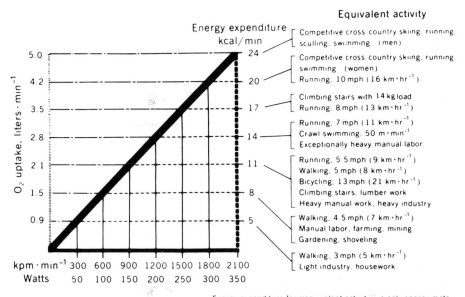

Figure 3.9. Approximate energy expenditure in different physical activities and equivalent cycle ergometer work (from Åstrand and Rodahl, 1986).

avoid fatigue and exhaustion and be able to leave the workplace with sufficient vigour left over for the enjoyment of leisure.

Work load

As already mentioned, work load may be expressed in terms of liters of oxygen uptake per minute. Examples of the energy cost of a variety of common everyday activities are given in Figure 3.9. In view of the fact that maximal aerobic power, i.e. maximal oxygen uptake, varies greatly from one person to another, a work load that is fairly easy for one worker may be quite exhausting for another. Suppose two men are to perform the same task, such as carrying a heavy load uphill, requiring an energy expenditure of about 2 liters of oxygen per min. One of the individuals has a maximal oxygen uptake of 6.0 liters $\cdot$ min^{-1}, the other 2.0 liters $\cdot$ min^{-1}. In the first case, the individual is merely taxing 30 per cent of his aerobic power. Consequently, he can probably carry on all day without fatigue, as is normally true when the work load is less than about 30–40 per cent of the individual's maximal aerobic power. Furthermore, he can continue to cover more than half his energy expenditure from the oxidation of fat, since his metabolic processes are essentially aerobic at this low relative work load. The second man, on the other hand, is taxing his aerobic power maximally and can carry on for only a few minutes, during which time he is compelled to rely on his carbohydrate stores as a major source of metabolic fuel because of the high relative work load.

Expressing the work load, as such, in absolute values (liter oxygen uptake per min) may thus be quite meaningless. Instead, it should be expressed as a percentage of the individual's maximal aerobic power. This means that the ratio between load and power should be assessed individually; that is, the individual's maximal oxygen uptake has to be determined and his or her rate of work has to be assessed. In general, the same principle also applies to the muscle groups which are engaged in the performance of the work in question, since only a certain percentage of the maximal muscle strength can be taxed without developing muscular fatigue.

With the development of mechanization, automation and many work–saving devices, modern technology has contributed greatly to the elimination of much heavy physical work. However, mechanization does not always lead to reduced energy expenditure. Thus Oberoi *et al.* (1983) measured the energy expended by hand washing clothes and by machine washing in 15 Indian women. The mean energy expenditure in washing clothes by hand while sitting on a 6.5 cm stool, and while standing at sink level were not significantly different from the mean energy expenditure found during washing clothes by machine.

The human being is not ideally suited to be a source of mechanical power. In this respect he cannot compete with modern mechanical devices, such as a bulldozer or a truck. The power output of an average man engaged in prolonged work over an 8-hour working day may amount to little more than 0.1 horsepower (1 horsepower = 750 watts). A horse may yield at least 7 times that amount, and an ordinary farm tractor, 70 horsepower.

In most instances, at least in the Western world with its advanced technology, prolonged excessively heavy work can easily be eliminated with technical aids; it is merely a matter of cost and priority. Establishing limits for permissible physical work loads is therefore of less practical value than considering the manner in which the work is being performed, the worker's opportunity to influence the working situation and to govern his rate of work, the safety and the general atmosphere of the working environment, and the arrangement of work shifts, etc. In most jobs in modern industry the worker or operator is able to adjust the rate of work according to personal capacity. However, there are some exceptions, as when the work is performed by a team. Here, the weak have to keep up with the strong. In such teamwork, older

Table 3.1. *Prolonged physical work, classified as to severity of work load and cardiovascular response*

In terms of oxygen uptake:	
Light work	up to 0.5 liter $\cdot$ min^{-1}
Moderate work	0.5–1.0 liter $\cdot$ min^{-1}
Heavy work	1.0–1.5 liter $\cdot$ min^{-1}
Very heavy work	1.5–2.0 liter $\cdot$ min^{-1}
Extremely heavy work	over 2.0 liter $\cdot$ min^{-1}
In terms of heart rate response:	
Light work	up to 90 beats $\cdot$ min^{-1}
Moderate work	90–110 beats $\cdot$ min^{-1}
Heavy work	110–130 beats $\cdot$ min^{-1}
Very heavy work	130–150 beats $\cdot$ min^{-1}
Extremely heavy work	150–170 beats $\cdot$ min^{-1}

workers, who are generally slower and who have a reduced physical working capacity, may be hard-pressed to keep pace with the younger members of the team. In any event, great individual differences do exist in physical working capacity, and practical experience has indicated that a work load taxing 30 to 40 per cent of the individual's maximal oxygen uptake is a reasonable average upper limit for physical work performed regularly over an 8-hour working day. Similarly, no more than 40 per cent of maximal muscle strength should be applied in repetitious muscular work in which the time of each muscular contraction is about one-half the time of each period of relaxation (Banister and Brown, 1968).

The physiological and psychological effects of a given energy output (per min, per 8-hr, per day) are determined by the individual's maximal aerobic power, the size of the engaged muscle mass, working position, whether work is intermittent at a high rate or continuous at a lower intensity, and environmental conditions. In general, a person's subjective experience of a particular work load or rate of work is more closely related to heart rate than to oxygen uptake during the performance of the work, since the work pulse, in addition to the actual work load, also reflects emotional factors, heat, the size of the engaged muscle groups, etc.

Bearing these reservations in mind, the following identification of prolonged physical work, classified as to severity of work load and to cardiovascular response, may be of some use. Table 3.1 refers to average individuals twenty to thirty years of age, and can only be used as general guidelines in view of the vast individual variations in ability to perform physical work.

Fatigue

Subjective feelings of fatigue usually occur at the end of an 8-hour workday when the average work load exceeds 30–40 per cent of the individual's maximal work capacity, depending on whether or not the work is more or less continuous or intermittent. These symptoms may range from a slight feeling of tiredness to complete exhaustion. Attempts have been made to relate these subjective feelings to objective physiological criteria such as the accumulation of lactate in the blood. While such a relationship is often observed in connection with prolonged strenuous physical efforts such as an athletic event, this relationship is not usually present in prolonged light or moderate work.

Christensen (1960) defines physical fatigue as a state of disturbed homeostasis due to work and to the working environment. It should be emphasized, however, that, so far, very little is known about the nature of this disturbed homeostasis. Thus, all that can be said with certainty at present is that the fall in the blood sugar observed in a fasting subject engaged in prolonged submaximal work lasting several hours causes disturbed homeostasis in the central nervous system leading to a feeling of fatigue as one of the symptoms of hypoglycemia. This hardly ever occurs in regular work. It is also clear that the accumulation of lactic acid in muscles engaged in intense work involving anaerobic metabolic conditions is a sign of disturbed homeostasis leading to symptoms of local fatigue. But this is not likely to occur in ordinary manual labour.

In her search for objective symptoms of fatigue, I. Åstrand (1960) observed a rise in heart rate in subjects working at a load corresponding to about 50 per cent of the individual's maximal oxygen uptake during a period of about 8 hours. However, since these experiments were carried out during the day (morning and afternoon), it is still an open question as to whether these changes are in fact due to, or partially due to, the development of fatigue, or the result of the normal circadian rhythms, which cause both a rise in heart rate and rectal temperature. In a normal environment, a rise in rectal temperature above 38 °C is noted when the oxygen uptake exceeds 50 per cent of the individual's maximal oxygen uptake (I. Åstrand, 1960).

Volle *et al.* (1979) examined the possible fatigue induced in workers submitted to a compressed work week, i.e. 40 hours in 4 days, as compared with the usual 40 hours in 5 days. Two groups of workers from two different factories manufacturing similar products, one group practicing the 4-day week, the second the 5-day week, participated in the study. The data did not reveal any significant difference between the two groups in terms of reaction time, heart rate, blood pressure, body temperature, O_2 uptake, CO_2 output, etc., measured before and after the first and the last day of the week. The only difference observed was that the critical flicker fusion frequency and the right-hand strength showed a significantly greater impairment in the 4-day week group. However, to what extent these parameters do in fact represent a greater level of fatigue may be questioned.

A study in a Norwegian machine shop failed to disclose any objective signs of fatigue, neither local nor general, at the end of a normal working shift (unpublished results).

Small or large muscle groups

The smaller the muscle mass involved in handling a certain load, the greater the stress on each individual muscle fiber. The heart rate is also higher in arm work than in leg work, especially in prolonged work (see Figure 3.10); it is particularly high when working with the arms above the head (I. Åstrand *et al.*, 1968). The arterial blood pressure is also significantly higher when working with the arms than when working with the legs at a given work load (oxygen uptake) or cardiac output (P.-O. Åstrand *et al.* 1965), indicating that arm work may represent a heavier load on the heart. Heavy work with the arms such as digging or shovelling snow may therefore be inadvisable for cardiac patients or for untrained older individuals. In addition, one is apt to hold one's breath while lifting the load, and this increases the intrathoracic pressure which, in turn, may hamper the normal venous return to the heart (the Valsalva effect). The maximal oxygen uptake in arm work is about 70 per cent of the maximal oxygen uptake attained when working with the legs.

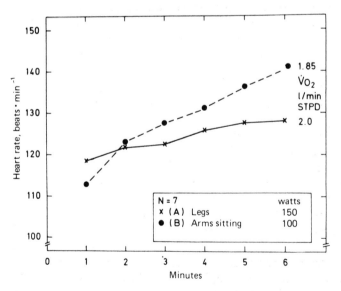

Figure 3.10. *Heart rates during the first 6 min of cycling at 900 kpm·min⁻¹ (150 watts) (A), and of arm cranking at 600 kpm·min⁻¹ (100 watts) (B) in a sitting position (modified from Vokac et al., 1975).*

Both in static and dynamic muscular work, endurance is related to the developed force expressed as a percentage of the maximal force which the muscle can develop (see Chapter 14, Figure 14.3). Thus, the stronger the muscles, the greater load they can endure without developing muscular fatigue. Malmquist *et al.* (1981) have shown that a stereotyped task has a greater tendency to produce localized muscle fatigue than a more varied one, even if the latter is heavier.

Work rhythm

With the development of mechanization, automation and many work-saving devices, modern technology has contributed greatly to the elimination of much heavy physical work. Nevertheless, occasionally some heavy physical work is still unavoidable in a number of occupations, such as commercial fishing, agriculture, forestry, construction, transportation and many service occupations. In some cases, when the work load appears excessively high, it is evident that the only reason why the task can be accomplished is that it is performed by intermittent work, i.e. brief work periods interspaced with short periods of rest. However, the tendency is for the elimination of physical strain, while at the same time the need for increased output and greater efficiency through rationalization and automation has resulted in an accelerated tempo in most industrial operations. The shorter working week has resulted, on the whole, in a greater work intensity. The outcome is increasing nervous tension and mounting emotional stress. Consequently, the greatest problem in many industrial operations today is not the physical load, but rather the mental stress and unfavourable working environment.

Continuous work

With increasing rate of work, there is a linear increase in oxygen uptake, up to the point when the oxygen uptake levels off, i.e. a further increase in work load is not accompanied by a further increase in oxygen uptake. The individual has reached his maximal oxygen uptake, or maximal aerobic power (Figure 3.11). At this stage the energy metabolism is largely anaerobic, and the lactate accumulation in the muscle tissue and in the blood rises sharply. The more motivated the subject the higher the concentration of the lactate level at the point when the subject has to stop.

At submaximal work loads, fed individuals could endure 6 hours of work (15-min rest per hour of exercise) at a load corresponding to approximately 60 per cent of their maximal oxygen uptake without

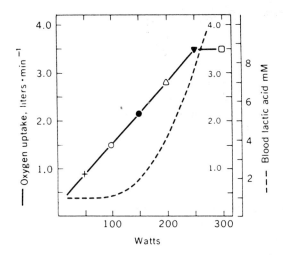

Figure 3.11. Oxygen uptake measured after 5 min of exercise, and plotted in relation to the rate of exercise, and peak lactate concentrations in the blood at the end of each exercise period (from Åstrand and Rodahl, 1986).

difficulty. When fasting, the same subjects could barely complete the work period at the same work load. The difference was due to the difference in the blood sugar level. In the first case, the blood sugar level remained more or less unchanged, while in the latter case the subjects had a marked drop in their blood sugar level (Rodahl *et al.*, 1964). In fasting subjects, at work rates up to 75 per cent of the individual's maximal oxygen uptake (15-min rest during each hour of work), and tolerated for perhaps as long as 3–6 hours, the hepatic glucose supply may be the limiting factor, causing a drop in the blood sugar. This leads to central nervous system symptoms typical of hypoglycemia (dizziness, nausea, confusion, blackout, etc.) without the depletion of muscle glycogen depots (Pruett, 1971).

During the recovery after physical work there is no sudden drop, but a gradual decline in oxygen uptake evidently associated with the energy cost of replenishing the TTP, PCr and oxygen stores of the body. This is followed by a much slower decline, which may last for a few hours, depending upon the severity and duration of the work, which may be explained as caused by such factors as the elevated body temperature, circulating catecholamines, etc. due to the preceding work stress.

Intermittent work

A physiological steady state work situation is rarely attained in industrial work, because ordinary muscular work is very seldom maintained at a steady rate for long periods. The typical muscular

activity encountered in industrial work or in recreational pursuits is
intermittent work.

In an experiment by I. Åstrand *et al.* (1960), a subject exercised on a
cycle ergometer at an extremely heavy intensity of 412 watts, producing about the same total amount of work (247 kJ or 25.200 kpm) in half
an hour by intermittent work, varying the length of work and rest
periods (Figure 3.12). When working continuously at this rate (412
watts) he was exhausted at the end of 3 min. When exercising intermittently for 1 min and resting for 2 min he was able to continue for 24
min before being completely exhausted, at which point his blood lactate
concentration had reached 15.7 mM. When reducing the exercise
periods to 3 seconds, interrupted by 60 seconds rest periods, his blood
lactate level remained below 70 mg $\cdot$ 100 ml^{-1}. When the work periods
were reduced to 10 seconds interspaced with 20 seconds rest, he was
able to accomplish the 247 kJ within 30 min with no severe feeling of
strain, with a blood lactate concentration of about 2 mM, or close to

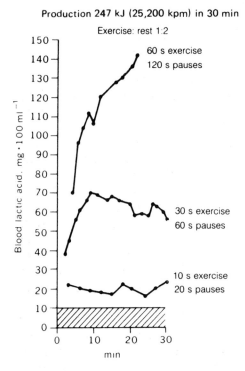

Figure 3.12. *Blood lactate concentration in a total work production of 247 kJ (25.200 kpm) in 30*
min intermittant work. The work rate was 412 watts (2520 kpm$\cdot$min^{-1}) and the
exercise period 10, 30 and 60 seconds, and rest periods 20, 60 and 120 seconds
respectively (from I. Åstrand et al., 1960).

resting level. The authors conclude that the critical factor is the length of the work periods; the duration of the rest periods and the total time spent resting during the 30 min period were only of secondary importance. Part of the explanation of the effect of the micropauses in intermittent work with the short work periods of 10 seconds may be the role played by the stored oxygen in the muscle tissue in the form of oxymyoglobin.

The practical conclusions to be drawn from these observations is that the heavier the work rate, the shorter should be the exercise periods. In fact, this may explain why older individuals, in spite of their reduced maximal aerobic power, are able to remain in jobs involving heavy manual labour, such as forestry, farming, construction work and service functions involving carrying heavy loads.

Static versus dynamic work

As pointed out in Chapter 1, man is made to work dynamically, changing between work and rest, between muscle contraction and relaxation in more or less a rhythmic fashion. Only a very small segment of our muscle mass is made to endure prolonged or lasting loads, such as our antigravity muscles.

In static work, the muscles involved may be more or less maximally contracted (if the work load is high enough), causing the intramuscular pressure to be higher than the blood pressure, obstructing the circulation of the blood to the region in question. This will interfere with the delivery of oxygen and substrates to the muscles involved, and removal of metabolites and carbon dioxide. Furthermore, the venous return of the blood from the extremities is hampered since the venous pumps, which depend on rhythmic contracting muscles, are non-functioning. Finally, intense static muscular contraction is often associated with breathholding (Valsalva manoeuvre), increasing the intrathoracic pressure, obstructing the venous return of blood to the heart. This reduces the venous filling of the heart, which in turn reduces the stroke volume; and in an endeavour to keep up the cardiac output, the heart rate will increase. The result of all this is a highly undesirable cardiovascular situation, which in young healthy individuals is easily tolerated, but which in cardiac patients and older individuals may have harmful or even fatal effects.

Less dramatic but equally devastating effects may result from more prolonged static work at a much lower load as encountered in many types of modern mechanical or electronic industries in which the worker for hours on end, day after day, is staring at a minute object, while manipulating fine instruments at an extremely high degree of precision. This entails the constant accommodation of the eye lens, i.e. tensing of the muscles adjusting the curvature of the lens; the fixing of the vision, involving the constant tensing of the external eye muscles; the fixing of the position of the head, involving the muscles of the head and neck; the fixing of the arms, involving the tensing of the arm- and shoulder muscles for prolonged periods. The consequences of this will be discussed in more detail in Chapter 14.

Work schedules

Man was evidently made to work during the day, and to rest and recuperate during the night. Otherwise he would probably have been provided with night vision as good as that of the owl. It is common knowledge that our performance capacity varies with the time of day, more so in some individuals than in others, and that some people function better in the early part of the day (Type A), while others function better later in the day (Type B). This, apparently, is linked to inherent circadian rhythms.

In humans, a number of physiological functions, such as heart rate, oxygen uptake, rectal temperature and urinary excretion of electrolytes and catecholamines, show distinct rhythmic changes in the course of a 24-hr period. The values are at their lowest during the night, and rise during the day, reaching their peak in the early afternoon. This phenomonen is known as circadian rhythms, and is thought to be regulated by several separately operating biological clocks. It occurs in most individuals, although there are apparently a few exceptions; some individuals show reversed rhythms, the rectal temperature, for example, being highest at night (Folk, 1974).

These rhythmic changes in physiological functions are, as mentioned, associated with changes in performance. This relationship appears to be especially obvious in the case of rectal temperatures and performance. In general, the lowest performance is observed early in the morning (about 4 a.m.). Thus, the delay in answering calls by switchboard operators on night shift was twice as long between 2 and 4 a.m. as during the daytime (Colquhoun, 1971). A similar relationship may exist also in the case of athletic performance. Thus, A. Rodahl et al. (1976), studying the performance of top swimmers who competed under comparable conditions early in the morning and late in the evening, found that the swimmers performed significantly better in the evening than in the morning ($p < 0.001$).

Shift work

The fact that human beings are 'day-animals' and that some of their basic physiological functions which are associated with their perform-ance capacities are subject to circadian rhythmic changes, suggests that humans may not ideally be suited for night work. Nonetheless, some work has to be done at night. Consequently, shift work has been practised for generations in one form or another. Yet, little precise information is available as to what effects shift work has on physiological functions or physical performance, and there is no general agreement as to what type of shift work or work schedule is to be preferred. Most of the available information refers to clinical, social or psychological aspects of shift work (Aanonsen, 1964). A review of the literature indicates that the health of shift workers in general is good in spite of such complaints as loss of sleep, disturbance of appetite and digestion, and a high rate of ulcers. The social and domestic effects of shift work represent greater problems than do the physiological effects. The results of studies pertaining to the effects on productivity are conflicting, as are results concerning accident rates. Absenteeism because of illness appears to be lower among shift workers than among day workers. It has been suggested that the physiological and biological effects are probably related to circadian rhythms rather than to work schedule. To what extent such circadian rhythms are related to health, performance and a feeling of well-being is still undetermined.

Systematic studies of men engaged in rotating shift work and in continuous night work (Vokac and Rodahl, 1974, 1975) indicate that shift work represents a physiological strain on the organism. It causes a desynchronization between functions such as body temperature and the biological clocks governing these functions. These studies show that there are considerable individual differences in the reaction to shift work, supporting the general experience that not everyone is equally suited for such work (some individuals consistently show relatively high values for urinary catecholamine elimination during shift work, whereas others have consistently low values). As judged by the catecholamine excretion, the greatest strain occurs when the worker, after several free days, starts work on night shift. The results of these studies indicate that it is preferable, from a physiological standpoint, to distribute the free days more evenly throughout the entire shift cycle; that is, to alternate between work and free time regularly instead of assigning several consecutive free days.

The study of continuous night work shows that at the onset, body temperature and work pulse fell in the course of the night as if the subject were sleeping, although he was working (Figure 8.1; Chapter 8). It takes several weeks for this normal rhythm to be reverted (i.e. for an increase in body temperature in the course of the night work). In view of this, it would appear unrealistic to keep shift workers on continuous night work for prolonged periods in order to obtain the benefit of the reverted physiological reactions, since such a reversion takes too long to occur and is lost when interrupted by a single day.

Disturbances in circadian rhythms may give rise to considerable problems for those who have to travel by air from one continent to

another in order to conduct business, to take part in political negotiations, or to participate in athletic competitions. It is an open question
whether the indisposition or functional disturbances experienced after
such intercontinental flights are in fact due to disturbed circadian
rhythms, to loss of sleep, or to both. It is a common experience,
however, that by being able to sleep during such travel, if necessary by
using sleep-producing drugs, the individual can maintain a reasonable
functional capacity in spite of the rapid shift from one time zone to
another. (For further details see Rutenfranz and Colquhoun, 1978 and
1979; and Chapter 8 in this volume).

Working position

Working in a standing position may represent a greater circulatory
strain than does working in a sitting position. Conversely, working in a
standing position, which will permit the worker to move about and
thereby vary the load on individual muscle groups and facilitate
circulation, may at times be preferable.

Working technique may be of major importance in conserving
energy and in providing varied use of different muscle groups. The
monotony of a working operation may be a stress for some individuals,
but a relief for others who can carry on the work more or less
automatically while thinking about something else.

The tempo of work performance may be extremely important and
impose stresses which, in some instances, may be unbearable or harmful
to the individual.

Environment

Physical performance may be influenced by the external environment,
both directly and indirectly. Air pollution may affect physical performance directly by increasing airway resistance and thereby pulmonary
ventilation, and indirectly by causing ill health. Noise may cause
hearing loss, but as a stress it may also cause an elevation of heart rate
and affect other physiological parameters which reduce physical performance. Low ambient temperature may in itself reduce physical
performance by causing numbness of the hands or lowered body
temperature. But it may also hamper performance because of the
hobbling effect of bulky clothing necessary for the protection against
cold and the slowing down of ordinary simple functions because of
snow and ice. Heat, if intense, may greatly reduce endurance because of
the need for more of the circulating blood volume to be devoted to the
transportation of heat rather than to the transportation of oxygen, and
because of the effect of dehydration which may accompany heat

exposure as the result of loss of body fluids through sweating. High gas pressures, as encountered in modern underwater operations, present a variety of rather unique problems for the work physiologist, while the impairment of physical work capacity at high altitudes is an old problem, and perhaps one of the best studied problems concerning the environmental effects on physical performance.

Temperature

The human being is a warm-blooded animal. Since both physical and mental performance, within limits, is temperature dependent, increasing with increasing temperature up to a point, the evolution of the human being has elected to attain this performance advantage at the expense of the cost of heating the human body to a temperature around $37\,^{\circ}C$ or higher. In the words of A.V. Hill (1939): 'At our temperature, which is much higher than that of a frog, our nerve messages go four to eight times as fast as his.' The fact that physical performance is temperature dependent is the reason why athletes warm up prior to athletic competitions. For a more comprehensive treatment of the physiological effects of environmental temperature, see Åstrand and Rodahl (1986), and Chapters 10 and 11 in this volume.

Cold

In subjects exposed to cold, an impairment of the performance of serial choice reaction time has been observed, as well as an impairment of manual dexterity (Wyon, 1982). The latter may, in part, be explained by changes in nervous conduction velocity. At a local skin or tissue temperature of $8-10\,^{\circ}C$, a complete nervous block may occur (Vangaard, 1982). Bergh (1980) found that physical performance is reduced with reducing core temperature. This temperature effect amounted to 4–6 per cent per $^{\circ}C$ in maximal exercise of less than 3 minute duration, and 8 per cent per $^{\circ}C$ at 3–8 minute exercise. He also observed reduced maximal dynamic muscle strength with reducing muscle temperature.

It appears that our temperature regulation is programmed to maintain the core temperature at the expense of the shell temperature, especially that of the hands and feet. Sudden cold exposure of the hand may cause an almost complete shutdown of the circulation to the exposed hand (Vangaard, 1975). Since manual dexterity is impaired with falling hand temperature, unprotected exposure to cold may lead to reduced manual performance. In addition, it may lead to an increased risk of accidents, as well as cold injury.

Exposure of the face to cold air not only leads to an increased peripheral vasoconstriction, but also to a slowing down of the heart rate

(Le Blanc *et al.*, 1978; Riggs *et al.*, 1981, 1983). In our own experiments (unpublished data), we have observed a sudden appearance of ST-segment depressions in the ECG in patients with coronary artery disease upon entering our cold chamber.

Heat

Both neuromuscular function and the energy-yielding processes, which are the basis for human performance, attain their optimal levels at temperatures above those typical of the resting state. At temperatures substantially higher than these optimal levels, both physical and mental performance may deteriorate due to an interplay of a number of factors. One of the contributing factors is circulation of the blood. In excessive heat the available circulating blood volume, generally speaking, has to serve a dual purpose: in addition to transporting oxygen it has to transport heat from the interior of the body to the skin where it can be dissipated in order to prevent overheating. This may in turn limit the oxygen transporting capacity of the blood circulation, since phylogenetically temperature regulation appears to have priority over oxygen transport (Hardy, 1967). Furthermore, hypohydration caused by prolonged heat stress in itself impairs performance, especially endurance. As already mentioned, prolonged heat strain may also impair mental and psychomotor functions, thus affecting performance.

For reasons explained above, heat exposure represents an extra load on the cardiovascular system. Consequently, exhaustion occurs much sooner during heavy physical work in the heat because the blood, as already mentioned, in addition to carrying oxygen, now also has to serve as a cooling fluid. To do this, while still maintaining the cardiac output, the heart rate has to increase. At the same time the circulation time is decreased. This obviously represents an extra burden on the heart to which hydrostatic factors in the case of prolonged standing work may be added.

Williams *et al.* (1967) observed an increase in heart rate and a fall in stroke volume at submaximal work rates in the heat. They also demonstrated a greater lactate accumulation in subjects working in the heat, compared to a neutral environment, probably due to the reduced muscle blood flow in the heat.

Heat stress is not only a major problem in the traditional hot industries, i.e. industries using processes in which energy is transferred into heat, such as those involved in production and casting of steel, iron, ferroalloys, aluminium, magnesium, siliconcarbide, etc. It is also a problem in industries which do not produce heat but where the operations are performed in an excessively hot environment, such as deep mining for gold or coal. Heat stress can also be a significant problem for pilots during low-level flights in hot climates, especially in fighter aircraft that impose high task loads, as pointed out by Nunneley and Flick (1981), who noted lowered G-tolerance and increased general fatigue in pilots on the hotter flights, and observed weight losses up to 2–3 per cent.

Any objective assessment of mental performance during exposure to heat (or cold) is hampered by the lack of suitable testing methods and large individual variations. It is generally agreed that the upper limit for optimal function is about 25°C for the unacclimatized clothed individual (Pepler, 1963). The deterioration in performance observed at temperatures higher than this refers to the precise manipulation requiring dexterity and co-ordination, the ability to observe irregular faint optical signs, the ability to remain alert and the ability to make quick decisions. Wyon *et al.* (1979) consider that the deterioration in mental performance observed by them in young individuals under conditions of moderate heat stress (up to 29°C) may be explained by reduced arousal in the absence of conscious effort.

The impairment of the physical performance capacity observed in individuals exposed to prolonged excessively hot environments may, to a considerable extent, be due to hypohydration or dehydration as the result of sweating. Irrespective of the cause of sweating (work or heat, or both), hypohydration is associated with a decrease in stroke volume and a concomitant increase in heart rate during submaximal work (Figure 3.13). Although maximal oxygen uptake, maximal cardiac output and maximal stroke volume are not modified by hypohydration up to 5 per cent of body weight, endurance, i.e. the time that a

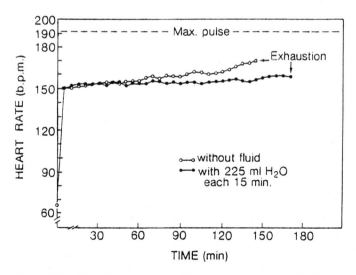

Figure 3.13. Mean heart rates in subjects running on the treadmill at 70 per cent maximal O_2 uptake until exhaustion, with and without fluid (from Staff and Nilsson, 1971).

standardized maximal work load can be tolerated, is definitely reduced after dehydration (Figure 3.14). The rectal temperature is also significantly higher in dehydrated subjects (Figure 3.15), the rise being related to the weight loss incurred (Gisolfi and Copping, 1974), and probably caused by inadequate sweating (Greenleaf and Castle, 1971).

Some degree of acclimatization to heat may be achieved in a week, by daily exposure to a hot environment for about an hour. Adequate acclimatization for a normal 6–8 hour shift, however, requires a minimum of 4 hours per day of heat exposure, combined with physical activity for at least 8–9 days (Wyndham *et al.*, 1973). Heat acclimatization is associated with an increased sweat production, a lowered skin temperature and a reduced heart rate (Figure 3.3). The basis for this acclimatization to heat is the fact that the sweat glands have the capacity to produce more sweat than they do under ordinary circumstances. The increased sweat rate after acclimatization provides for a more effective cooling of the skin through the evaporative heat loss, and the lowered skin temperature in turn provides for a better cooling of the blood flowing through it. Thus, the body can afford to cut down on the skin blood flow.

The available evidence suggests that heat tolerance is reduced in older individuals (Leithead and Lind, 1964). They start to sweat later than do young individuals. It takes longer for their body temperature to return to normal levels after heat exposure.

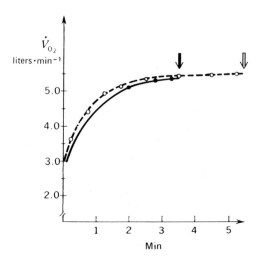

Figure 3.14. Oxygen uptake at an exercise load which could be tolerated for 5½ minutes during normal conditions (open symbols), but only for 3½ minutes after hypohydration (filled symbols). Arrows indicate maximal exercise time (from Saltin, 1964).

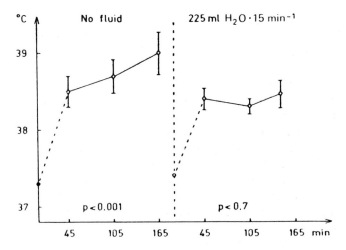

Figure 3.15. Mean rectal temperature in subjects running on the treadmill at 70 per cent maximal O₂ uptake, with and without fluid. When no fluid was taken, rectal temperature rose significantly from the 45th to the 165th minute (from Staff and Nilsson, 1971).

Women, having a higher tissue conductance in heat than do men, require lower evaporative cooling in both hot-wet and hot-dry environments (Shapiro *et al.*, 1981). This fact indicates a greater variation in the peripheral reaction to climatic stress in women, although this fact, apparently, is of no practical importance for the performance of work.

Humidity

The moisture content of the ambient air may affect performance when working in a hot environment, especially if the work load is heavy, causing profuse sweating. At ambient temperatures above the mean skin temperature, the only possible way of losing heat to the environment is by evaporative cooling, i.e. by the evaporation of sweat from the skin surface. The more saturated the air with water vapour, the less likely it is for the sweat to evaporate, i.e. the more limited is the evaporative cooling. The sweat starts to drop off the skin when the sweat intensity has reached about one-third of the maximal evaporative capacity (Kerslake, 1963). This excess sweat represents a useless loss of body fluid. This is a matter of considerable practical importance in many industrial work places, emphasizing the importance of adequate fluid intake.

The combination of moderate cold and high humidity may be experienced subjectively as unpleasant, but carefully controlled experimental data as to the physiological effect of such conditions of wet-cold are lacking. From the available evidence it appears that moderate wet-cold is of minor physiological consequence, although the effect of dampness on the insulative value of the protective clothing may be a matter of some concern.

Air velocity

The motion of the air surrounding the working individual may be a factor enhancing the cooling in hot conditions, by increasing evaporative cooling and by removing the layer of warm air molecules surrounding the individual under cold conditions. For this reason this factor has to be taken into account when assessing the thermal stress of the environment by unitary indices, as is done in the case of the Botsball or Wet Globe index.

The effect of air velocity is particularly apparent when the air is cold and chilly, especially as far as the exposed parts of the body are concerned. This wind chill factor was incorporated into the concept of a Wind-Chill Index by Paul Siple and his associates during an expedition to the Antarctic to show the rate of cooling of a heated body exposed to the wind (Siple and Passel, 1945; Court, 1948). In spite of the limitations of this approach (Burton and Edholm, 1955), tables of wind-chill values are quite useful as a practical guide to cold tolerance.

Reduced gas pressure (high altitude)

Human beings are ideally suited to live on the surface of the earth, preferably close to sea level. Under optimal conditions, i.e. during periods when the barometric pressure is high, man may venture as high as to the top of the world's highest mountain, Mount Everest, without supplementary oxygen (Sutton *et al.*, 1983). Yet, permanent human habitation is not encountered above an altitude of 4,500 m. The air pressure in pressurized aeroplane cabins usually corresponds to an altitude of 1,000 to 1,500 m, which is enough to cause a reduced capacity to perform prolonged physical work involving large muscle groups. The reduced oxygen pressure in the inspired air at high altitudes initially causes a reduced maximal oxygen uptake (= reduced maximal work capacity), for which the various physiological responses cannot fully compensate. Consequently, the importance of the anaerobic energy yield increases.

A prolonged stay at high altitude causes, within a few weeks, an acclimatization to reduced oxygen pressure in the inspired air, and a long-term adaptation at the end of years at high altitude. These changes, which are reversible, cause a gradual improvement in physical performance and endurance at the altitude in question. (For a more detailed treatment of this subject, see Åstrand and Rodahl, 1986.)

High gas pressure

Increased pressure, as encountered in underwater operations in connection with modern offshore oil exploration, not only limits physical performance, but also imposes possible major health problems. When breathing air at a hyperbaric pressure during heavy physical work, the increased air density reduces the pulmonary ventilation and the maximal oxygen uptake may be reduced. However, by substituting nitrogen with helium which has a lower density, the pulmonary ventilation increases and the maximal oxygen uptake may in fact be greater than that measured in the same person during normal atmospheric conditions (Fagraeus, 1974).

While humans may become acclimatized to low air pressures (high altitude), there is no way to become biologically acclimatized to high air pressures, such as those encountered in deep sea diving, or when a submarine crew has to escape from the inside of the submarine (where the pressure is normal) to the surface through the sea (where the air pressure is much higher). The body may tolerate high gas pressures as long as the pressure is the same inside as it is outside the body. When using a snorkel connected to the mouth when diving, the atmospheric pressure is maintained in the lungs, while the surface of the chest is exposed to the water pressure in addition to the atmospheric pressure. At a depth of roughly one meter, at which point an additional pressure of 1/10 atm is acting upon the body, the pressure difference becomes so large that the inspiratory muscles no longer have the strength to overcome the external pressure, and normal breathing becomes impossible. For this reason a snorkel system does not permit diving to depths exceeding about one meter. At greater depths, a breathing apparatus has to be used, in which the pressure in the system corresponds to that prevailing at the depth in question. An overpressure in the system may cause lung tissue damage. As the pressure increases with increasing depth, more gases are taken up and dissolved in the various body tissues. At a depth of 10 meters an additional pressure of one atm acts upon the diver's body, and twice as much gas will be dissolved in the blood and tissue as at the sea surface. This is apt to give the diver trouble (the bends), mainly because of the nitrogen. (For a review of the problems of hyberbaric physiology, see Lambertsen 1967; Fagraeus, 1974; Behnke, 1978; Halsey, 1982; Lawrie, 1985, and Flook, 1987.)

Noise

Noise may affect performance by causing distraction and by interfering with the ability to concentrate; it may interfere with communication and may in severe cases even contribute to hearing loss. Noise with great variability in level, intermittent noise, high-level repeated noise, and noise with frequencies above about 2,000 Hz, is likely to affect performance. On the other hand, rhythmic noise may improve performance, especially in the case of simple, repetitive tasks.

The likelihood of developing hearing loss depends on the intensity, frequency spectrum and duration of the noise exposure, and to some extent on individual susceptibility. Loss of high-frequency components usually come first. Generally speaking, prolonged exposure to noise levels in excess of 85 decibels on the A scale of a sound level meter may lead to hearing loss. (For references see Rodgers and Eggleton, 1983. For a description of hearing protection programmes, see NIOSH, 1972).

Studies of the effect of background music in factories to improve the working environment have given varying results (Grandjean, 1961). However, there appears to be no conclusive proof that background music increases productivity, although most, but not all, production workers are likely to enjoy hearing music when they work, particularly in the case of repetitive assembly tasks. In jobs where more concentrated attention is required, music may be an undesirable distraction.

Vibration

Vibration may also affect human performance. It is usually characterized by its frequency, acceleration and direction. It may affect the whole body (whole-body vibration), or it may affect parts of it, such as the hands (segmental vibration). In the case of whole-body vibration, low frequency vibration (4–8 Hz) is particularly undesirable. According to Rodgers and Eggleton (1983), vibration is slightly better tolerated while standing than while sitting, and women, on the whole, may experience greater discomfort than men when exposed to the same level of vibration. It is the combination of frequency and acceleration that determines the level of discomfort. Vibration tolerance is usually defined as the level of vibration which causes a performance-impairment of a motor or visual task.

Vibrations with frequencies in the 0.1–20 Hz range, and accelerations generally less than 0.2 g, predominantly vertical, are common in most construction equipment, trucks and buses (1 g = approximately $10 \text{ m} \cdot \text{sec}^{-2}$). Exposure to this degree of whole body vibration continuously for more than 8 hours may be associated with considerable discomfort and objective physiological reactions. Vibration in the 2–20 Hz range at 1 g acceleration may cause abdominal pain, nausea, dizziness, muscle tension, chest pain and shortness of breath (for references, see Rodgers and Eggleton, 1983).

On the basis of a review of the literature concerning duration effects of whole-body vibration on comfort, Kjellberg and Wickström (1985) concluded that the time dependency proposed by the International

Standard for whole-body vibration, ISO 2631, appears to constitute an overestimation of the importance of exposure time for the strength of the effects.

Segmented vibration, especially of the hands, when using power tools, will vary according to the type of power tool being used, in that the weight, size and design of the tool will determine its vibration characteristics (Abrams and Suggs, 1977). As far as segmental vibrations of the hands are concerned, frequencies in the range of 25–150 Hz, and accelerations from 1.5 g to 80 g, are most likely to lead to problems such as Raynaud's Phenomenon, or white fingers, i.e. circulatory disturbances of the fingers causing numbness, pain and loss of strength (Ramsey, 1975). The patient may be unable to perform fine, manipulative tasks, particularly in cold environments.

Exposure to local vibration may cause a decrease in skin temperature (Iwata, 1974) associated with a reduced blood supply to the fingers (Makarenko, 1969), as well as an increase in plasma norepinephrine and epinephrine concentration (Krasavina *et al.*, 1977; Miyashita *et al.*, 1983).

Infrasound

Claims that infrasound (acoustic frequencies below approximately 16 Hz) may have an adverse affect upon human performance appear to have been exaggerated. Intensity levels higher than 120 dB of infrasound had no adverse effects on reaction time in humans. (For references, see Harris *et al.*, 1976.)

Magnetic fields

Possible performance effects and health hazards associated with exposure to steady and time-varying magnetic fields have been reviewed by Miller (1987). He emphasizes the dangers which may be encountered by people with sicklecell anemia or with medical electronic implants, such as pacemakers.

According to Knave and Floderus (1988) there is no solid evidence from current experimental studies suggesting an association between exposure to low-frequency electromagnetic fields and any type of cancer.

Air pollution

Occupational health aspects of air pollution (dust or airborne particles, or gases such as CO, CO_2, O_3, and SO_2), are beyond the scope of this book. Here we shall merely deal with certain aspects of the physiological implementations of some of these problems. This may include systemic effects, notably the effects on the respiratory system, such as the effect of tobacco smoking on airway resistance, which will briefly be dealt with as an example of a problem area where work physiologists may be of some use to the toxicologists and the occupational physician, because of their familiarity with available methods for the assessment of pulmonary functions.

Carbon monoxide (CO)

Carbon monoxide is a colourless, noxious gas, slightly heavier than air. It is formed by incomplete combustion of carbon-containing material, such as coal or coke, or by a number of chemical processes. Carbon monoxide, in the ambient air, stems mainly from the exhaust from combustion engines, heating plants and a variety of industrial processes, etc. Automobile exhaust contains up to 7 per cent CO, which in closed spaces, such as a garage, may be lethal within 5–10 min.

Cigarette smoke often contains enough CO to produce a blood carbon monoxide concentration of 3–10 per cent, corresponding to the inhalation of ambient air containing 20–50 ppm (parts per million) CO, which is approaching the recommended maximal permissable limits for CO exposure (Scandinavia: 35 ppm; Great Britain: 50 ppm). The correlation between the CO content of the inspired air, duration of exposure and the concentration of CO Hb in the blood is given in Figure 15.1 (from Peterson and Stewart, 1970). See also Peterson (1970); and Stewart *et al.* (1970).

Tobacco smoking

In view of the fact that tobacco smoking, and especially the smoking of factory-made cigarettes, may have a profound effect on an individual's maximal aerobic capacity, i.e. his ability to perform prolonged heavy physical work, a brief summary of the main facts may be pertinent in this connection. For a more extensive treatment of this subject, the reader is advised to consult Larson *et al.* (1961).

As in the case of all kinds of incomplete combustion, a certain amount of carbon monoxide is formed when tobacco is burned, whether it is smoked in the form of cigarettes, pipes or cigars. In fact, tobacco smoke may contain as much as 4 volume per cent of carbon monoxide (CO). Some of this carbon monoxide is absorbed by the smoker, especially if he inhales. But even without inhaling, some of the CO is absorbed, as is the case with individuals situated close to the smoker, inhaling some of the air containing tobacco smoke (passive smoking).

Carbon monoxide combines with the hemoglobin of the blood with an affinity some 200 times greater than that of oxygen. Consequently, any inhaled CO is likely to be combined with hemoglobin forming carboxyhemoglobin (COHb) at the expense of oxyhemoglobin. Thus the presence of even small amounts of carbon monoxide may reduce the oxygen-transporting capacity of the blood. Subjects smoking 10–12 cigarettes a day may have as much as 5 per cent carbon monoxide hemoglobin in their blood, those smoking 15–25 cigarettes a day over 6 per cent, and those smoking 30–40 cigarettes a day over 9 per cent (Larson *et al.*, 1961). At rest, carbon monoxide concentrations in the blood below 10 per cent give no subjective symptoms, the adverse effect is noticeable only during physical exertion. It should be borne in mind, however, that a blocking of 5 per cent of the hemoglobin by carbon monoxide will reduce the maximal oxygen uptake noticeably (Ekblom and Huot, 1972).

Comroe (1966) has shown that inhalation of the smoke from a cigarette will, within seconds, cause a 2–3 fold increase in airway resistance which may last 10–30 minutes. This acute effect is not caused by carbon monoxide or nicotine, but by particles in the cigarette smoke that have a smaller diameter than 1 μm (micrometer) and which affect the sensory receptors of the airway path. The chronic effect of tobacco smoking, on the other hand, is a swelling and an increased secretion in the mucous membrane of the respiratory tract, causing a narrowing of the air passages, leading to an increased airway resistance. This increased airway resistance is not noticeable at rest, when the pulmonary ventilation is less than 10 liters per min, but when the demand on respiration is elevated the increased respiratory resistance caused by smoking may be noticeable.

In carefully controlled experiments, Wahren *et al.* (1983), have shown that cigarette smoking significantly increases energy expenditure (measured by oxygen uptake). This may partly explain why smokers often gain weight when giving up smoking.

Chapter 4
Methods for the assessment of physical work capacity and physical work load

Introduction

In Chapter 3 we reviewed some of the factors which may affect our work capacity. In this chapter we shall turn our attention to:

1. How to assess this capacity objectively.
2. How to assess the work load.
3. How to assess the total stress of work, and how the individual worker reacts to it, i.e. the strain of work.

As is evident from Figure 3.4 in Chapter 3, the physical work capacity or maximal aerobic power (maximal oxygen uptake) varies greatly from one person to the next. It also varies with age, and it is usually higher in men than in women. Consequently, a work load which may be experienced as fairly easy by one person may be quite exhausting for another. Since it is the ratio between work load and work capacity that counts, the expression of the work load, as such, in absolute values, such as liters of oxygen uptake per min, may be quite meaningless. Instead, it should be expressed as a percentage of the individual's maximal aerobic power, i.e. how much of the individual's maximal aerobic power has to be taxed in order to accomplish the work in question. Ideally, therefore, the individual's maximal oxygen uptake should be determined, and the work load should be similarly assessed individually.

The same principle, matching load against capacity, also holds in the case of loads on individual muscles or small muscle groups. Thus, only a certain percentage of the maximal strength of a muscle can be taxed for any length of time without developing muscular fatigue in dynamic as well as in static muscle contraction (for references see Åstrand and Rodahl, 1986).

It should be emphasized that an individual's maximal oxygen uptake, measured during a three to six minute laboratory test, and which reflects the individual's ability to take up and to deliver oxygen to the working muscles under conditions of maximal or supermaximal loads, is not necessarily a measure of that individual's endurance in sustained, heavy physical work. Thus, in dehydrated subjects, the maximal oxygen uptake may not be reduced, but the time which the subject can maintain the maximal load is usually significantly shortened (Saltin, 1964).

Assessment of an individual's maximal work capacity (maximal aerobic power)

This can be done by direct measurement of the individual's maximal oxygen uptake, or it can be estimated from data obtained at submaximal tests.

1. Direct measurement of maximal oxygen uptake

The measurement of maximal oxygen uptake can be accomplished in the laboratory by treadmill running, by using a cycle ergometer or by a step test. For a critical review of tests of maximal oxygen uptake, see Shephard (1984). Generally speaking, treadmill running may give slightly higher values for maximal oxygen uptake than does ergometer cycling. The main disadvantage with the step test is that the subject's body weight may affect the results. For a more detailed discussion, see Åstrand and Rodahl (1986).

In treadmill running the subject is compelled either to follow the speed of the treadmill, or to jump off. This is not so in the case of the cycle ergometer; here the subject, in most types of ergometers, may continue to pedal at a reduced rate. Thus, the level of motivation is especially important in the case of ergometer cycling. Another disadvantage with the cycle ergometer is the strain on the subject's knees and legs at heavy work loads, sometimes forcing the subject to discontinue the exercise before the maximal oxygen uptake has been reached.

While most individuals know how to run, not everyone is familiar with bicycle riding. If not, the cycle ergometer may be an unsuitable instrument for the assessment of maximal oxygen uptake, as in the case of native Eskimos, for instance. It has been claimed that the maximal oxygen uptake of Eskimos, still living in their traditional primitive manner, is surprisingly low. However, a study was conducted with eight active Eskimo hunters, 30–51 years of age, on the north-west coast of Greenland in 1973. None of these Eskimos had ever used an ordinary bicycle before. Consequently, it was necessary to spend a considerable amount of time teaching them and training them how to use the pedals properly. In addition, an Eskimo drummer was employed to help the subjects keep the pedalling rate at the set number of revolutions per minute. The result was a mean maximal oxygen uptake of 53 ml O_2 per kg body weight, which compares favourably with results from our own population (Vokac and Rodahl, 1976). In four of these subjects the maximal oxygen uptake had been determined four years earlier by another investigator. Our figures for VO_2 max were about 50 per cent higher than those previously reported in the same subjects. This example may serve to emphasize the importance of the subjects being thoroughly familiar with the test method to be used.

Regardless of the type of maximal work test used, it should be preceded by a proper warming-up activity. The actual test may then be performed in one of several ways:

1. The subject may start to exercise at a load which, on the basis of preliminary submaximal tests, has been judged to be maximal or slightly supermaximal for this subject. The subject is encouraged to maintain this work rate for a period of 3 to 6 minutes.
2. The subject may be exposed to one or two submaximal work loads of five to six minutes duration, preferably interspaced with a rest period of some five minute duration, finally followed by a maximal or supermaximal work load. If a very accurate determination of the subject's maximal oxygen uptake is required, it may be necessary to conduct several such tests on different days, until the highest attainable oxygen uptake for this particular subject has been reached, i.e. the oxygen uptake has levelled off.
3. The work load may be increased stepwise at intervals (every minute, every other minute or every third minute) until complete exhaustion.

At best, the determination of an individual's maximal oxygen uptake is very difficult, requiring a considerable amount of experience. By definition, the maximal oxygen uptake is the highest oxygen uptake attainable in the subject in question, i.e. a further increase in work load will not result in an increase in oxygen uptake. This is not always easy to achieve. In most ordinary subjects, when the maximal oxgyen uptake is reached, the blood lactate level is usually about 8 mM (blood sample collected from the finger tip within one minute after the end of exercise) and the Respiratory-Quotient (R) is more than 1.1.

When it comes to the type of test to be used for the applied purposes under consideration here, the cycle ergometer has a number of practical advantages. Since the subject is seated on the ergometer, the mechanical efficiency is independent of body weight, and the energy expenditure can be predicted with great accuracy. The cycle ergometer is inexpensive, it can easily be transported and moved from one place to another, and in the case of a mechanically-braked cycle ergometer, it does not require electrical power. It should be kept in mind, however, that the oxygen uptake varies to a certain extent with the pedal frequency. Pedal frequencies between 40 and 50 rpm produce the lowest oxygen uptake (the greatest mechanical efficiency). Usually a pedal frequency of 50 or 60 rpm is preferred (60 in the case of a very heavy work rate) timed by a metronome. If the subject's respiratory rate is of interest, it should also be kept in mind that the subject's respiration tends to be affected by the pedal frequency.

Cycle ergometers which produce a constant power regardless of pedal frequency have the disadvantage at very high loads that if the subject is unable to keep up the tempo, the force required to move the pedals becomes so large that the subject is unable to move the pedals and is forced to stop prematurely, or before the investigator has completed his measurement or the collection of his air samples.

A practical guide for a quick assessment of maximal oxygen uptake, using an ordinary mechanically-braked cycle ergometer is given in Chapter 8 of Åstrand and Rodahl (1986). It is suggested that the test should start with a warming-up period, at a relatively low initial load, and to increase the load gradually until a heart rate of about 140 is attained (120 in subjects over 50 years of age). On the basis of this steady state heart rate and the power setting of the ergometer, the maximal oxygen uptake is predicted from the Åstrand and Åstrand nomogram (Figure 4.1). A work load on the ergometer is then selected which will correspond to an oxygen uptake about 10 to 20 per cent higher than that which was predicted as a maximum for the subject from Figure 4.1. If it should become apparent at the end of the first minute or so of the exercise that the subject is having difficulty in keeping up the pedalling rate, and starts to hyperventilate, the work load is slightly reduced so as to allow the subject to continue for a total of about three minutes. Or, if the subject, after a minute or two, should appear fitter than predicted, the work load may be slightly increased, providing the subject is still able to continue long enough for the investigator to be able to collect two Douglas bags of expired air, which will take at least one minute. The investigator's dilemma is usually to

Figure 4.1. Nomogram for the calculation of maximal oxygen uptake from submaximal heart rate using a cycle ergometer or a steptest. VO_2 max values obtained by drawing a line between the 'body weight' step test scale, or the 'work rate' ergometer scale to the heart rate scale. The predicted maximal O_2 uptake is read where the line crosses the middle scale. (from I. Astrand, 1960).

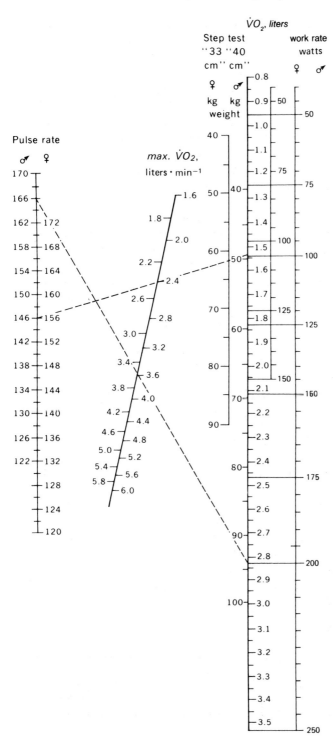

$\dot{V}O_2$, *liters*

Pulse rate

max. $\dot{V}O_2$,
liters · min⁻¹

Step test
''33 ''40
cm'' cm''

work rate
watts

decide when to put the noseclip and the mouthpiece with the respiratory valve in place on the subject, and to start the collection of expired
air. When in doubt, it is always better to start too early than too late.

The treadmill, as an instrument for producing a measurable and
repeatable work load, has the disadvantage that it is more expensive,
much heavier and therefore less movable than a cycle ergometer, and it
requires electrical power. Furthermore, the subject's body weight will
affect energy expenditure when walking or running on a treadmill
(Figure 4.2). Finally, the mechanical efficiency in the same subject may
be quite different when walking or when running on the treadmill,
which may cause some difficulties.

At any rate, it is advisable to allow time for the subject to become
accustomed to treadmill running, including how to jump off and on the
running treadmill safely and with ease. The use of a handrail for support
should be avoided since it is apt to make the metabolic demand more
unpredictable (Figure 4.3).

There are a number of different protocols in use for treadmill testing
of maximal oxygen uptake. In some cases it is a matter of (a) keeping the
speed of the belt constant, and to increase the work load by increasing
the slope (The Balke test); (b) keeping the inclination constant, but
increasing the work load by increasing the speed of the belt; or (c)
increasing the work load by increasing both speed and slope (the Bruce
protocol). Another method (d) is first to predict the subject's maximal

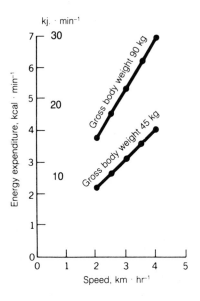

*Figure 4.2. Effect of speed and gross body weight on energy expenditure when walking
(data from Passmore and Durnin, 1955).*

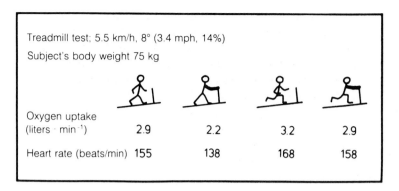

Treadmill test; 5.5 km/h, 8° (3.4 mph, 14%)

Subject's body weight 75 kg

Oxygen uptake (liters · min⁻¹)	2.9	2.2	3.2	2.9
Heart rate (beats/min)	155	138	168	158

Figure 4.3. The effect of using a handrail for support in a treadmill test
(from P-O. Åstrand, 1982).

rate of exercise on the basis of the heart rate at a submaximal rate of exercise on a treadmill or cycle ergometer, and then to select a setting of the treadmill speed and inclination on this basis as a starting point, keeping the speed constant, but increasing the incline every third minute until complete exhaustion (Saltin and Åstrand, 1967). For further details, see Chapter 8 in Åstrand and Rodahl, (1986).

A very simple method of assessing a person's state of physical fitness or maximal aerobic power is by having the subject stepping up and down on a step or a box at a fixed rate. This method has been extensively applied in field work, especially in military personnel. Even though the height of the step may be adjustable according to the height of the subject (Hettinger and Rodahl, 1960b), it has its limitations, however, not only because body weight may affect the results, but also because of the fact that at maximal effort the stepping pace has to be rather high, exposing the subject to the risk of stumbling.

In conclusion, then, it would appear that the most practical method for measuring maximal oxygen uptake in individual subjects in the field is a method based on the cycle ergometer. In reality, however, the determination of the individual's maximal aerobic power by direct measurement may not be strictly necessary except for academic purposes, since, in most cases, data of satisfactory accuracy for practical purposes may more easily be obtained by indirect assessment of maximal aerobic power, based on submaximal tests.

In the case of work stress–ECG testing of individuals with known or suspected coronary disease, it may be preferable to apply a near-maximal cycle ergometer test, using an electrically-braked cycle ergometer, starting at 100 W, and increasing the load stepwise by 25 or 50 W at three minute intervals. In this manner the test may be continued until exhaustion, or until the subject is exhibiting severe dyspnea, ECG signs (ST-depression 3 mm or more, or significant arrhythmias), a drop in systolic pressure by 10 per cent or more from the last measured value, systolic blood pressure over 300 mm Hg, musculoskeletal symptoms or other specific symptoms or signs necessitating cessation of the test. The ECG should be continuously monitored on an oscilloscope, and blood pressure should be recorded preferably at the beginning and end of each load. For a more detailed description of the exercise ECG test see Erikssen *et al.* (1977), and Mundal *et al.* (1987).

An individual's oxygen uptake is measured by analyzing the oxygen content of the air the subject is breathing in, and of the air he is breathing out, the difference being the amount of oxygen he has taken up. This means that the amount of air which has passed the lungs has to be determined, and a sample of this expired air has to be analyzed with respect to O_2 and CO_2.

For the purpose of collecting expired air, the Douglas bag method is still the method of choice, because of its simplicity, accuracy, and applicability under a wide range of circumstances. The method has the disadvantage that it can only provide an average figure for the oxygen uptake during the collection period, but this is quite acceptable for the purpose considered here. Another disadvantage is the fact that the subject is limited in his movements, since the bag is connected to the subject. (For a detailed description of the use of the Douglas bag method for the determination of oxygen uptake both in the laboratory and in the field, see Chapters 7 and 8 in Åstrand and Rodahl, 1986.) The room temperature during the test should not exceed $18-20\,^{\circ}C$, and the relative humidity should be kept below 60 per cent. By connecting two bags with the aid of a four-way stopcock, continuous air samples may be collected. At least 50 liters of expired air should be collected in the bag before the stopcock is turned back to room air or to another bag, the stopcock always being turned *during inspiration*.

A balanced spirometer is used to measure the volume of the expired air sample. The composition of the air sample, i.e. the O_2 and CO_2 content, is analyzed by manometric methods (Haldane or micro-Scholander) or by electric gas analyzers. For further details see Åstrand and Rodahl (1986).

2. Indirect assessment of maximal work capacity (maximal oxygen uptake) on the basis of submaximal tests

In most occupations in industrialized countries the work load to which the worker is exposed taxes a fairly modest fraction of his maximal aerobic power, usually less than 30 per cent. Only in occupations such as commercial fishing do work loads tax more than 40 per cent of the individual's maximal oxygen uptake regularly. In healthy individuals work loads of less than 25 per cent of the individual's maximal aerobic power are usually not enough to cause symptoms or signs of general physiological fatigue. In such cases measuring the worker's maximal aerobic power may not be called for. Yet there are instances when such an assessment would be of interest, especially when there is a need for evaluating the effect of changing work routines or improving work procedures, in which case an assessment of the work capacity-work load ratio might be useful in evaluating the effect of the changes made in terms of work load on the worker. In such cases, however, as is evident from the previous discussion, the direct measurement of maximal oxygen uptake is too complicated, time-consuming and requires too much in terms of technical facilities, equipment and qualified technical assistance to make this approach feasible. Furthermore, one may be reluctant to subject the individuals to be tested to maximal work loads, especially if they are not all young, fit and healthy. In addition, experience has taught us that even highly trained and technically well-equipped investigators may find it quite difficult to arrive at a precise figure for maximal oxygen uptake, in the case of untrained, middle-aged or older individuals not accustomed to such test procedures. As a matter of fact, it is not uncommon in such cases that each test session may in fact represent a training session, sufficient to produce a training effect, causing steadily increasing values for the VO_2 max from one test session to the next. In such cases it may be difficult to know when to stop; in any case it is very time consuming.

It is obvious, therefore, that in such cases, when there is a need to establish how much of a worker's manual aerobic power he is taxing in the performance of his work, or in the performance of a particular task, the indirect assessment based on submaximal work loads is the method to use.

Practically speaking, there are basically two different approaches available to us: one is based on the circulatory response to a work load, as evidenced by the heart rate response; the other is based on the participation of the anaerobic metabolic processes in the energy metabolism, as evidenced by the changes in the Respiratory Quotient (R) due to the production of lactic acid in the working muscles, causing hyperventilation and an excess pulmonary elimination of CO_2.

The assessment of maximal oxygen uptake on the basis of heart rate response to submaximal work loads: The Åstrand-Åstrand nomogram (Åstrand and Ryhming, 1954)

In sustained physical work the energy transformed into mechanical work by the muscles, largely by aerobic processes, is directly related to the amount of oxygen taken up by the lungs, and transported by the blood circulation to the working muscle cells. The basis for this oxygen delivery is the amount of blood which the heart can pump per minute, i.e. the cardiac output. The cardiac output is in turn a product of stroke volume times heart rate. While the amount that the stroke volume can increase with increasing demand is rather limited (less than 50 per cent), the heart rate may increase from a resting rate of say 70 beats per minute to a maximal rate of say 200 or even higher. Thus, the maximal heart rate may be a limiting factor in an individual's maximal oxygen transport capacity; or more precisely, the difference between maximal and resting heart rate is proportional to the range of increase in oxygen transport from rest to maximal exertion. This is the basis for using the heart rate at specific submaximal work loads to predict an individual's maximal work capacity.

From Figure 4.4 it is clear that with increasing work load there is a linear increase in heart rate. If the line in Figure 4.4 is extended to the rate which is the maximal heart rate for this particular individual, say 200 beats per min, his maximal oxygen uptake may be predicted to be 4.5 l per min, by following the vertical line from the 200 heart rate mark. This is the justification for using the heart rate at submaximal work loads to predict the individual's maximal oxygen uptake. The problem is, however, that an individual's maximal heart rate cannot be accurately predicted because it varies greatly from one individual to another. Of two individuals in the same age group, one may have a maximal heart rate of less than 180, the other perhaps 220. It is therefore not justifiable to use this approach to predict maximal work capacity in different individuals for comparison. But it is justified

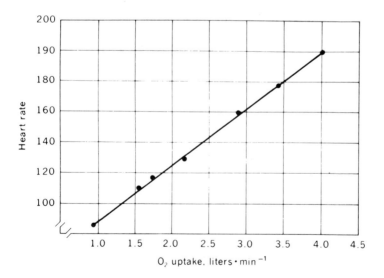

Figure 4.4. *The relationship between heart rate and O_2 uptake on the cycle ergometer (from Åstrand and Rodahl, 1986).*

when using the same subject as his own control for comparison, for instance before and after a training program, or to compare two different work situations in one and the same subject, serving as his own control, since the maximal heart rate of this subject is the same. Thus, it is quite acceptable to use this indirect approach for the assessment of the effect of altered working procedures or working routines in terms of how much of the worker's maximal aerobic capacity is being taxed before and after the change has been introduced.

The best known, and the most widely used submaximal test for the prediction of maximal oxygen uptake, is the one based on the nomogram of Åstrand and Ryhming (1954) discussed in more detail in I. Åstrand (1960). This nomogram is constructed on the basis of data from 86 male and female physical education students, who were relatively well-trained. Basically, the data were the results of cycle ergometer experiments, but the nomogram can also be used in connection with a step test, for the purpose of predicting maximal oxygen uptake on the basis of heart rate, at a known work load. At a work load 50 per cent of the maximal oxygen uptake, the mean heart rate was 128 beats per min in the men, and 138 in the women. At 70 per cent of their maximal aerobic power, it was 154 for the males and 164 for the females. This was the actual basis for the construction of the nomogram (Figure 4.1). Thus, when constructing the nomogram, the individual's maximal heart rate was not taken into consideration. It was noted, however, that the predicted oxygen uptake of individuals over some 30 years of age in most cases was overestimated. The reason for this is the fact that maximal heart rate declines with age. A correction factor for age was therefore introduced, as set out in Table 4.1.

The large scatter of the heart rate at a given oxygen uptake due to the great individual differences was noted by the initiators of the nomogram. Thus, at an oxygen uptake of 3.0 liters per min, heart rates varied

Table 4.1. Maximal heart rate decline and age (from I. Åstrand, 1960)

Age	Factor
15	1.10
25	1.00
35	0.87
40	0.83
45	0.78
50	0.75
55	0.71
60	0.68
65	0.65

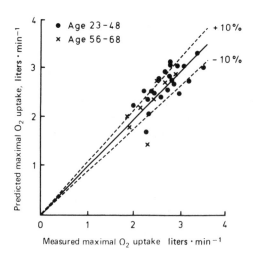

Figure 4.5. The relationship between estimated maximal O₂ uptake calculated from the Åstrand nomogram, and actually measured maximal O₂ uptake in 22 male subjects, ages 23 to 48 (•), and 9 male subjects, ages 56 to 68 (x). Broken lines denote a deviation of ± 10 per cent from the 'ideal line' (from Rodahl and Issekutz, 1962).

from 140 to 220. The standard error of the prediction was 10 per cent in well-trained individuals of similar age, but up to 15 per cent in moderately trained individuals of different ages, even though the age correction factor was used. On the whole, untrained individuals are often underestimated, while well-trained athletes are often overestimated (Åstrand and Rodahl, 1986).

The consequence of a standard error of 15 per cent is that of a group of 100 subjects, for whom the oxygen uptake is predicted to be 3.0 liters per min, 5 may actually have a maximal oxygen uptake less than 2.1, and 5 higher than 3.9 liters per min. This is the inherent limitation of any test such as this, which attempts to predict maximal levels of fitness on the basis of recorded parameters at submaximal cardiopulmonary loads. As a matter of fact, this test is as good as any, but even so, it may be improved by some minor modification, as suggested by Siconolfi *et al.* (1982). Nonetheless, the nomogram test has survived the test of time. It has repeatedly been critically evaluated, comparing the predicted results according to the nomogram with actually measured maximal oxygen uptakes in a number of subjects and under various conditions, and found reasonably acceptable when considering mean values for groups of people (Figure 4.5), or when used to assess the effect of training etc. in one and the same individual.

Prediction of the aerobic work capacity on the basis of the Respiratory Quotient (R)

In individuals who are taking certain cardiac drugs, such as β-blockers, which affect the heart rate, or who have cardiac arrythmia, the heart rate obviously cannot be used for the prediction of maximal aerobic work capacity. In such cases the maximal oxygen uptake can be predicted from the measurement of the Respiratory Quotient (R), i.e. the ratio CO_2/O_2 during a single 5-min cycle ergometer test according to a method developed by Issekutz *et al.* (1962). This is based on their observation that ΔR (delta R, i.e. exercise R minus 0.75) increases logarithmically with the rate of work. In most cases the maximal oxygen uptake is reached at a ΔR value of 0.40. This method is based on the fact that the accumulation of lactate in the blood during brief exercise sufficiently strenuous to activate anaerobic metabolic processes will stimulate the respiratory centre to cause the subject to hyperventilate, causing an excess of CO_2 to be exhaled. The CO_2/O_2 ratio (R) will increase gradually, reaching a value of 1.1 or higher when maximal oxygen uptake is reached. Blood lactate concentration at this point is usually in excess of 8 mM.

Assessment of work load

The physical work load to which the worker is exposed can be objectively assessed either by (a) direct measurement of the energy expenditure required to do the job, i.e. determination of the worker's oxygen uptake during the performance of the work in question, by collecting and analyzing samples of his expired air, or (b) indirectly by recording the worker's heart rate during the performance of the work.

Of these two approaches, obviously the *direct measurement of the oxygen uptake* of the individual worker while carrying out the actual work in question is the most accurate method. The disadvantage with this approach, however, is that the measured oxygen uptake only represents the energy expenditure for the time period during which the expired air sample is collected (usually a matter of minutes), which may not be representative for this work operation in general, or for the whole working day. Furthermore, the equipment used for the collection of the expired air may restrict or affect the worker, causing the test situation to be atypical, and may even hamper the actual work operations. In addition, the subject may object to the use of the noseclip and the mouthpiece, necessary for the expired air collection.

The *indirect assessment* of the work load, on the basis of the continuously recorded heart rate, reveals, on the other hand, a general

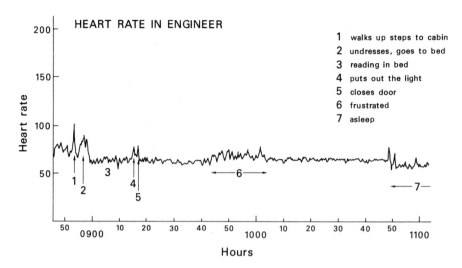

Figure 4.6. Heart rate of an engineer off duty on board a container ship: (1) he is walking up the stairs to his cabin; (2) he undresses; (3) reading in bed; (4) puts out the light; (5) gets out of bed to close a slamming door and returns to bed, but is unable to sleep because of a problem which has been annoying him. Then suddenly, at 10:50 his heart rate drops to less than 60 beats·min⁻¹, indicating that he has fallen asleep.

picture of the activity level during the entire working day, both in general, and during specific time periods. Moreover, by this method, it is possible to relate the individual worker's reaction to different work operations as judged by the heart rate response, by keeping detailed time–activity records for each subject, showing the precise timing of the different activities, which can then be indicated on the heart rate graphs (see Figure 4.6 as an example). For these reasons, the indirect assessment of work load based on the recorded heart rate may be preferable in many cases.

Direct measurement of oxygen uptake during the actual work operation by indirect calorimetry

The indirect calorimetry approach is based on the established validity of using oxygen uptake as a basis for measuring energy expenditure. This is accomplished by collecting samples of expired air during the performance of specific work operations, and analyzing the O_2 (and CO_2) content of the collected samples by conventional methods in a field laboratory. With the highly portable devices for the collection of expired air now available, and the rapid methods for analyzing the oxygen and carbon dioxide content of the air samples, it is now possible to carry out rather sophisticated and extensive studies involving the measurement of energy expenditure which reflect actual work loads in most kinds of work situations, including industrial work. A large body of data of this kind pertaining to the energy cost of physical work is already available (see Åstrand and Rodahl, 1986). Figure 3.9, Chapter 3, presents the energy cost of a number of common everyday activities based on measured oxygen uptake, compared with equivalent loads on the cycle ergometer.

The classical method is to collect the expired air in a Douglas bag carried on the subject's back, and to measure the volume of the collected air samples by a conventional spirometer in a field laboratory nearby (for details pertaining to the Douglas bag method, see Åstrand and Rodahl, 1986). However, instruments are now available which are carried on the subject, and which measure the volume of the expired air with the aid of flow meters. Aliquot samples of the expired air are collected at intervals, usually in small rubber bladders. One of the earliest devices of this kind is the Max Planck Respirometer (Müller and Franz, 1952). Another example is the integrating motor pneumo-tachygraph, or the IMP (Wolff, 1958). The Oxylog produced by P.K. Morgon Ltd., Chatham, Kent, United Kingdom, is a small battery-operated instrument which is carried on the subject's back; it is capable of measuring pulmonary ventilation and oxygen uptake during a wide range of submaximal work loads (0.3–3.0 l O_2 per min). It consists of a face mask with respiratory valves connected by an air hose to a flowmeter, communicating with a computer carried by a shoulder harness or a waist belt. The expired air passes through a mixing unit. With the aid of a miniature pump, samples are drawn for analysis in a polarographic oxygen sensor. Another sensor measures the oxygen partial pressure in the ambient air. On this basis the computer calculates the oxygen uptake, correcting for temperature and barometric pressure, assuming a relative humidity of 50 per cent in the inspired air; the results are immediately available by digital display (Harrison *et al.*, 1982; Louhevaara *et al.*, 1985).

In the cases when the oxygen content of the expired air is not electronically analyzed with the aid of sensors, as in the Oxylog, the collected samples of expired air are analyzed with respect to O_2 and CO_2 content by conventional gas analysis techniques in the field, either by volumetric methods (Haldane or Micro-Scholander techniques), or by physical methods such as those based on infrared, paramagnetic or mass spectrometric principles. If an accuracy of roughly ± 10 per cent is acceptable, it may be sufficient to analyze only the oxygen content of the expired air sample with the aid of a portable O_2 analyzer. For further details see Consolazio *et al.* (1963), Durnin and Passmore (1967), Banister and Brown (1968), Morehouse (1972), Kamon (1974), Harrison *et al.* (1982), and Wilson and Sklenka (1983).

Indirect assessment of energy expenditure (oxygen uptake) on the basis of recorded heart rate

Occasionally it may be of interest to assess the severity of a specific work load in absolute values, in terms of the required energy expenditure, i.e. oxygen uptake. As a rule, however, the purpose of assessing the work load to which an individual is exposed is to determine how much of the individual's work capacity or aerobic power he is taxing in order to accomplish the particular task in question.

When a subject is made to exercise on a cycle ergometer at a given steady work load for five to six minutes, and the work load is increased by say 25 or 50 watts in a stepwise fashion every five to six minutes, while the heart rate is being recorded at intervals, say every minute, the heart rate increases in a linear fashion with the increase in work load (see Figure 4.7). Since a given submaximal work load on the cycle ergometer always demands a given oxygen uptake, the mechanical efficiency being constant, there is also a linear relationship between heart rate and oxygen uptake, providing there are no other interfering factors such as emotions etc. affecting the heart rate. Thus, the above mentioned linear relationship between work load and heart rate also holds for oxygen uptake and heart rate (Figure 4.4).

Therefore, if the work load–heart rate interrelationship is established for a given person on the cycle ergometer, this linear relationship can also be used to determine the oxygen uptake equivalent of a given work

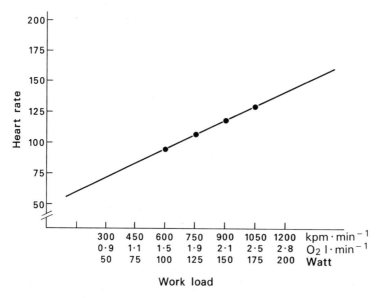

Figure 4.7. The linear increase in heart rate with increasing work load on the cycle ergometer.

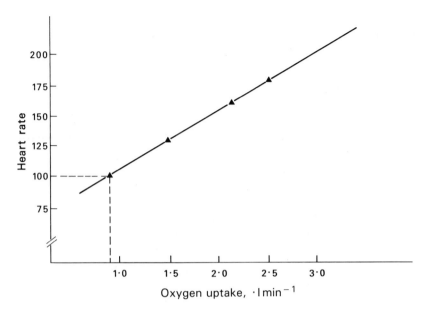

Figure 4.8. Oxygen uptake equivalents of recorded heart rates in a 21-year old coastal fisherman. In this case, a recorded heart rate of 100 beats·min⁻¹ is equivalent to an oxygen uptake of about 0.9 l·min⁻¹.

operation, if the heart rate during that particular work operation is recorded, without actually having to measure the oxygen uptake. An example of this is presented in Figure 4.8, which represents a 21-year old coastal fisherman in whom the heart rate was recorded while he was engaged in net fishing at the Norwegian fishing grounds off Lofoten. It is observed that in this case a recorded heart rate of 100 beats per minute represents an oxygen uptake of about 0.9 l per min.

This heart rate-oxygen uptake relationship is only valid when roughly the same large muscle groups are engaged in the work in question as are used in working on the cycle ergometer. It is also essential that the environmental temperature and the state of emotional stress are the same. Furthermore, the relationship is not valid when working with the arms overhead, as in the case of hammering nails into a ceiling, in which case the heart rate is exceptionally high (I. Åstrand *et al.*, 1968).

It should be borne in mind that the estimation of oxygen uptake from recorded heart rate may be subject to considerable inaccuracy. However, Rodahl *et al.* (1974), made 24 direct measurements of oxygen uptake with the Douglas bag method and compared the results with the oxygen uptake calculated from the simultaneously recorded heart rate in six commercial fishermen while engaged in coastal fishing. The difference was less than ±15 per cent, as is evident from Figure 4.9.

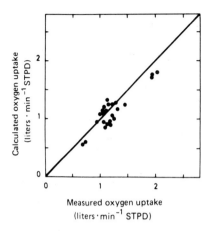

Figure 4.9. *Relationship between O₂ uptake measured by the Douglas bag method and calculated from the simultaneously recorded heart rate (from Rodahl et al., 1974).*

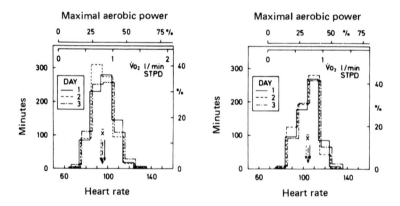

Figure 4.10. *Weighted means (x) and distributed curves of the heart rate and corresponding work load in two subjects during three consecutive days of observation. Left panel: 41-year old long-line fisherman; right panel: 56-year old catch handler (from Rodahl et al., 1974).*

The reproducibility of the results from such field studies, estimating work load on the basis of the recorded heart rate, is surprisingly good. Thus, Rodahl *et al.* (1974) examined the results from day-to-day over a period of three consecutive days in a fisherman out at sea, and in a catch handler working on land. They observed a remarkable reproducibility of the day-to-day results in the same individual doing the same work, as is evident from Figure 4.10. The weighted mean heart rates were practically the same for all three days, and the distribution curves, when superimposed, are very similar.

As previously pointed out, both in this chapter and in Chapter 3, the heart rate is considerably higher in arm work than in leg work at the same work load. Although this is indeed the case in prolonged work lasting 5 minutes or more, Vokac *et al.* (1975) have shown that the difference in heart rate between leg and arm work at the same work load is not very marked at the onset of the work, but increases as the work proceeds (see Figure 3.10, Chapter 3). In view of the fact that most ordinary work operations involve a dynamic type of work with rhythmic alternation between muscular contraction and relaxation, in which each period of work effort is rather brief, it appears that the use of the recorded heart rate as a basis for the estimation of work load may be acceptable even in many work situations involving arm work, or the use of small muscle groups.

It is thus clear that the recorded heart rate in the field, compared with the heart rate at known work loads on the cycle ergometer, may be used as a basis for the estimation of the work load, especially when the work in question involves the use of the same large muscle groups as are used in the cycle ergometer work.

A rough estimate of the work load in individual cases may be obtained without establishing the heart rate–work load relationship on a cycle ergometer, simply by recording the heart rate during the performance of the task in question. The heart rate as such may then be

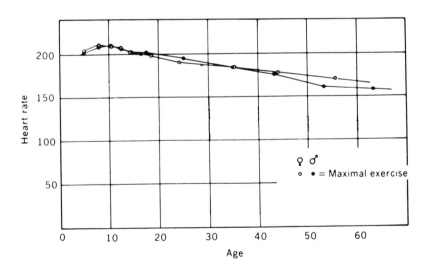

Figure 4.11. The decline in maximal heart rate with age. Mean values from 350 subjects. The standard deviation in maximal heart rate is about ± 10 beats/min in all age groups (from Åstrand and Christensen, 1964).

used as a rough expression of work load, or it can be expressed as a percentage of the estimated maximal heart rate. A rough estimation of maximal heart rate in an individual of known age may be derived from Figure 4.11, or even more roughly by subtracting the individual's age in years from the number 220.

In most individuals, a heart rate around 120–130 beats·min^{-1} corresponds to a work load of about 50 per cent of the individual's maximal oxygen uptake. Incidentally, this is also the work load when most individuals begin to be slightly out of breath.

Heart rate reserve

The extent to which a person may increase his work rate depends in part on how much he can increase his heart rate from the resting level to his maximal level, because the increase in heart rate plays a major role in increasing the cardiac output from rest to maximal work. The potential for an increase in heart rate from rest to maximal work, a parameter which may be termed 'heart rate reserve', is therefore an important indication of an individual's aerobic power potential. One way of expressing the circulatory strain which a given work load imposes on a subject is to express the heart rate at the given work load as a percentage of the heart rate reserve of that particular individual (the heart rate reserve being the difference between maximal heart rate and resting heart rate):

$$\frac{\text{HR work} - \text{HR rest}}{\text{HR max} - \text{HR rest}} \times 100 = \text{HR reserve}$$

In order to overcome the time-dependent variability, the heart rate results may be presented as weighted means (X_w), which are calculated from the individual values multiplied by the fraction of the sum of their duration (Rodahl *et al.*, 1974).

Ambulatory recording of heart rate

The recording of heart rate in the field is most conveniently accomplished with the aid of portable, miniature, battery-operated recorders or loggers, of which there are several versions on the market. In the past we have made extensive use of electromagnetic tape recorders such as those produced by Avionics in Philadelphia, USA, or Hellige in Freiburg/Breisgau, Germany, or Oxford Instruments in England. They work on the principle of monitoring the heart rate continuously in the form of a simplified electrocardiogram. The continuous recording or logging of the heart rate can also be accomplished by the use of a portable miniature computer, such as the Vitalog, produced by the Vitalog Corporation in Palo Alto, California, USA. The recorded heart rate coupled with time–activity records shows the degree and variation of the circulatory strain experienced by the subject engaged in different kinds of work operations. The heart rate graphs, thus obtained, may be replayed and transcribed on recording paper and graphically displayed and evaluated (I. Åstrand *et al.*, 1973), or evaluated by computer analysis (Rodahl *et al.*, (1974). A more recent and most promising development in the field of ambulatory monitoring is the Squirrel meter/logger, made by Grant Instruments Ltd., in Cambridge, England (Figure 4.12). It is a combined meter and data logger, i.e. it can be used to measure and immediately display data such as heart rate, tempera- ture, etc. At the same time it can store the recorded data in memory for immediate or later transfer to a small battery-operated portable PC, to a battery-operated portable printer in the field, or to a conventional office PC for data analysis, printout or graphic display. The 1200 series Squirrel is powered by six 1.5 volt alkaline batteries. The unit is designed specifically to use as little power as possible in the logging mode; the batteries can be expected to last for months. With care, the batteries can be changed without loss of stored readings. The unit also displays an indication of the remaining battery life. It is robust and extremely easy to operate. It has a number of channels, allowing several parameters, including heart rate, to be recorded or logged at the same time. This permits the investigators to record environmental param- eters, such as ambient temperature etc., and the subject's reaction to this

Figure 4.12. The Squirrel meter/logger.

temperature in terms of core and shell temperature, as well as heart rate changes. Another advantage with this unit is that it can display the data as they are being recorded, while at the same time storing the readings in the memory. It is also able to display readings stored in its memory. The unit is controlled by a microprocessor, which is programmed to perform all the required measurements, display and storage operations. An interface socket is fitted, which enables stored readings to be piped through a separate computer such as an Epson HX20 or Epson PX8 for display and storage. Software programs are also available for direct data transfer to IBM-compatible personal computers, for subsequent data analysis, display and printout.

The more recent 12-bit Squirrel is equipped with separate channels for the counting of the heart rate, based on the principle of counting the number of R-waves per minute. It is supplied with filters for the exclusion of artifacts or interference, and in controlled laboratory tests has been found to be quite accurate and reliable. It weighs about 1 kg and can be carried in a leather carrying case by a shoulder strap or by a belt around the waist. For heart rate recording the three disposable precordial electrodes are stuck to the skin and connected to a separate powerpack-converter unit connected with the Squirrel. The use of the Squirrel for the purpose of recording and logging other parameters than heart rate, and especially the logging of several different parameters simultaneously, will be discussed later.

In summary, continuous recording of the heart rate permits an uninterrupted collection of data which reflect the subject's work load during the entire work shift, or during the entire work day. A quantitative numerical analysis of the recorded data paired with detailed activity logs, and supplemented by visual analysis of the replayed heart rate graphs, permits a comprehensive and dynamic evaluation of the circulatory strain imposed by work loads of varying intensity. The use of loggers and computers make it possible to analyze large series of observations and ambulatory recordings of environmental stresses and physiological reactions to them. It also enables the observer to assess the relative severity of the work load, and the occurrence and duration of work loads exceeding a certain percentage of the individual's maximal aerobic power, or how much of the individual's heart rate reserve (the maximal heart rate minus the resting heart rate) the person is taxing in the performance of certain tasks or work operations.

For practical purposes, the heart rate, as such, may also be used to indicate the severity of the work load. A heart rate of 130 in most individuals roughly corresponds to about 50 per cent of the individual's maximal aerobic power.

The total stress of work

As is evident from the previous discussion in this chapter, the stress of work in terms of physical work load can be objectively measured and precisely expressed in numerical terms. The same applies to environmental stresses such as temperature. But the combination of several stressors, including mental stress, cannot be precisely measured and uniformly expressed. The total stress of work may, however, to a considerable extent, be assessed by the physiological strain it imposes on the working organism. The severity of the total stress, physical as well as mental, imposed on the working organism by a given work situation, is generally reflected by the magnitude of the organism's combined nervous and hormonal response to that stress. The magnitude of this response, from all indications, is more or less proportional to the severity of the stress.

Sensory inputs which may be interpreted by the organism as a challenge to its integrity, in a stereotyped manner, automatically trigger the mobilization of its physical resources in readiness for coping with the emergency, more or less in accordance with the fight-or-flight reaction conceived by Walter B. Cannon. According to Cannon's diary (Herd, 1972), he realized that the basic purpose of the autonomous hypothalamic stress reaction is to put the organism into readiness for whatever muscular activity the situation may call for, and that failure to obtain a normal release for this state of organic tension in modern man might, in the long run, have pathological effects.

This trigger mechanism or the neuro-hormonal 'alarm' reaction is initiated by groups of cells in the hypothalamus which react on the sensory conception of a challenge by releasing neuropeptides, or releasing factors, which are transported by the portal system to the anterior segment of the hypophysis. Here they cause the instantaneous production and release into the blood stream of hormones capable of producing major physiological alterations, all of which serve to enhance the organism's physiological capacity to perform muscular work. The growth hormone (somatotropin) causes an immediate release of free fatty acids from the adipose tissue into the blood stream, thus making it available as fuel for the working muscles. The adrenocorticotrophic hormone ACTH causes the release of cortisol in the adrenal cortex, which, via the blood stream, has a profound effect on almost every cell in the body. In the liver, proteins are broken down by cortisol to glucose, causing an increase in blood glucose levels, thus enhancing the fuel supply for muscular work, as well as fuel for the brain cells which normally depend on glucose as their metabolic fuel. Protein-resynthesis is largely blocked by cortisol, cells in the renal tubuli are stimulated to preserve sodium ions, and thereby also preserve the body water. Indirectly, cortisol may affect the smooth muscle fibers in the small arteries and veins, increasing their epinephrine-norepinephrine sensitivity, thus causing contraction and hence reduced blood perfusion in non-vital organs, such as the skin, gastro-intestinal tract and kidneys, for the benefit of organs essential for performance and survival, and which now may receive a major share of the circulating blood volume. Furthermore, the performance of the cardiac muscle fibers and the cellular membranes are enhanced.

The simultaneous activation of the sympathetic nervous system causes the release of epinephrine from the adrenal medulla. The stimulation of the nerve endings in the sympathetic nerve fibers causes the release of norepinephrine, which causes the mobilization of free fatty acids (FFA) to be used as fuel for the muscular work which is anticipated. Epinephrine causes the release of glucose from glycogen in the liver, and to some extent also of FFA from the adipose tissue, to serve as fuel for muscular work. Epinephrine also causes an increase in the heart rate, as well as an increase in the force of contraction of the heart.

Parallel with all these basic physiological alterations in preparation for an anticipated muscular effort, a release of physiological pain-killers, i.e. neuropeptides, such as encephalin which have an analgesic effect, may take place in the brain, and which make the organism less sensitive to pain.

As a result of all these profound physiological stress reactions there is an immediate increase in heart rate and pulmonary ventilation (both frequency and depth), obviously for the purpose of increasing oxygen uptake and cardiac output. A major proportion of the circulating blood volume is transferred from the guts and skin to the muscles where it will be needed for the muscular activity to follow. There is a major increase in the liberation of fuel, both glucose and free fatty acids, to be used in the anticipated muscular work. All these profound cellular alterations serve to enhance the exchange of metabolites, to facilitate energy metabolism, to preserve body water, and to prevent pain. The obvious principle behind all these changes is that those cells which are most active also receive most of the support, in terms of oxygen and fuel.

All these changes are merely the result of a trigger mechanism, aimed at preparing the organism to be ready for a muscular effort. During the actual physical activity that follows, the autonomous regulatory mechanisms automatically take over, adjusting the physiological state of the body in accordance with the local and general needs dictated by the activity itself. This also applies to the transition to rest and recuperation following the period of activity.

From this brief description of the physiological consequences of the total stress of work, both physical and mental, it is evident that the physiological response to the stress, i.e. the strain, may in principle be measured in terms of different indications of autonomous nervous system reactions (such as heart rate), or in terms of hormonal release (such as the level of epinephrine and/or norepinephrine in the blood or in the urine).

Hormonal response

Since the total stress response of the organism is reflected by the sympatheticoadrenomedullary activity, this response may be roughly assessed by measuring the level of epinephrine and norepinephrine (catecholamines) in the blood, or the amount eliminated in the urine. As the collection of blood samples from workers at their actual places of work may involve a number of practical problems, apart from the fact that the actual venepuncture and the drawing of the blood samples in itself may cause a stress-induced elevation of catecholamine levels in the blood, the measurement of urinary excretion of epinephrine and norepinephrine may be preferable. This may be done by the method described by von Euler and Lishajko (1961), and modified by Andersson *et al.* (1974), using the resting night urine as a base value. The urinary catecholamine elimination, expressed in ng·min^{-1} may be used as a measure of occupational stress (Levi, 1968). A nearly ten-fold increase in the epinephrine excretion and about a four-fold increase in norepinephrine excretion were observed during the workday, compared with the excretion during the night in coastal fishermen (I. Åstrand *et al.*, 1973). Even greater increases in norepinephrine excretion were observed in War College cadets during a strenuous 5-day simulated battle course, including strenuous physical activity, reduced food intake and sleep deprivation (Figure 4.13).

An increase in the urinary elimination of both epinephrine and norepinephrine may be caused by a number of factors, individually or combined. Circadian rhythm (Fröberg *et al.*, 1972), and changing from recumbent to standing position (Sundin, 1956), increase the catecholamine excretion during the day. It may increase markedly during physical exertion (von Euler and Hellner, 1952), in the cold (Lamke *et al.*, 1972), and due to emotional factors (von Euler, 1964, Levi, 1967).

There is no doubt that urinary catecholamine excretion may be used to quantitate stress response in laboratory experiments, and in stimulated work situations (Cox *et al.*, 1982). It may also be used in field studies, providing the collection of urine samples is carefully supervised and adequately controlled. The fact remains, however, that it is at best very difficult to obtain a complete urinary collection. The volume of residual urine is a source of error, and it is difficult to obtain samples more frequently than every two hours or so. This means that acute stress situations of brief duration may not be detected in the urinary sample, which shows the average value for as much as a couple of hours. In addition, there are considerable individual differences in urinary catecholamine elimination, even among persons exposed to the same work stress (Figure 4.13). Furthermore, extensive unpublished studies of urinary catecholamine elimination in a large number of workers at a variety of work places, which were carried out by

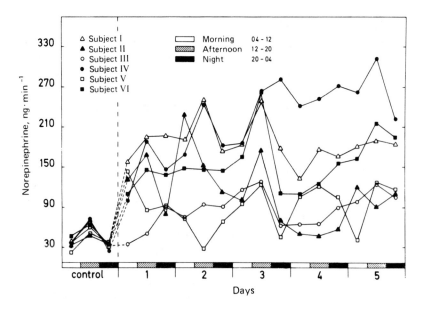

Figure 4.13. Urinary excretion of norepinephrine in War Academy cadets during simulated battle course (from Holmboe et al., 1975).

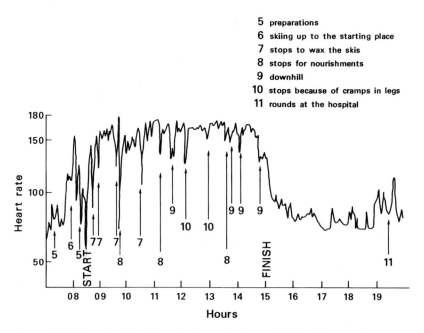

Figure 4.14. Heart rate in a 54-year old participant in a 70-km ski race.

members of the Norwegian Institute of Work Physiology, have failed to show any significant elevation of excretion rates in most industries. For these reasons this method may not be recommendable for the assessment of ordinary occupational work stress. Instead, continuously recorded heart rate may preferably be used as an indication of autonomous nervous system response to stress, both physical and mental.

Nervous response

An increased sympathicotonus, brought about by mental as well as physical stress, will give rise to an accelerated heart rate, which consequently may serve as an index of stress response. As already explained, the heart rate increases linearly with increasing physical work load, provided there is no major change in the subject's emotional state (see Figure 4.7). It should also be kept in mind that the heart rate may vary markedly as a consequence of emotional stress in an individual who is at rest or subject to a constant light work load, as shown in Chapter 6. The fact that the mean heart rate increases during demanding mental loads was shown by Hitchen *et al.* (1980), who observed that the heart rate increased by as much as 15 per cent during a demanding mental task.

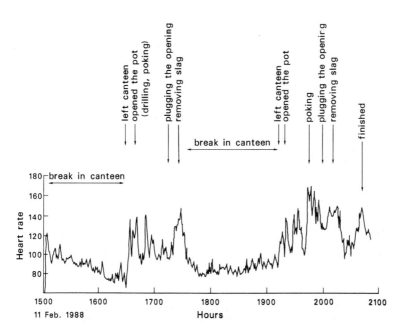

Figure 4.15. Heart rate in a worker in a ferromanganese production plant showing the effect of periods of strenuous physical work in the heat, interspaced with prolonged break periods in the canteen.

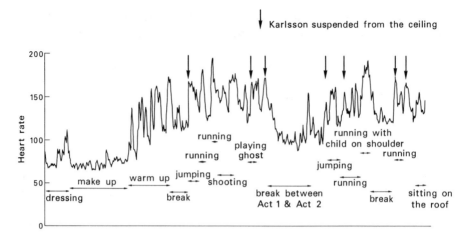

Figure 4.16. The heart rate reaction to a combination of physical and mental stress in an actor, acting part of the time suspended by a rope from the ceiling, pretending to be flying, in the play Karlsson on the Roof.

Combined with a detailed activity log kept by the subject himself (provided he is properly instructed), the continuously recorded heart rate may indeed provide useful information concerning a subject's response to stressful situations. Examples of heart rate reactions to physical stress are given in Figure 4.14, which shows the heart rate of a participant in a 70-km ski race, and in Figure 4.15, which shows the heart rate of a worker in a ferromanganese production plant during intense work periods followed by prolonged breaks. Examples of heart rate reactions to mental and emotional stress are given in Chapter 6. An example of the reaction to combined physical and mental stress is given in Figure 4.16, which shows the heart rate of a young actor playing the role of Karlsson in the children's play *Karlsson on the Roof*. Part of the time he is suspended by a rope from the ceiling, pretending to hover about in the air while continuously gesticulating and talking very loudly.

Heart rate variability

The degree of variation in a person's heart rate has been suggested as an indicator for non-physical stress. The basis for this is that there is a reduced tendency for fluctuations or variability in the heart rate when a person is subject to a mental load (Hitchen *et al.*, 1980). However, in a critical evaluation of this idea, Luczak (1979) points out that while this method is a good indicator of non-physical strain in the laboratory with steady states and well-defined and controlled conditions, it has limited value in the evaluation of strain in practical situations with varying environmental, physical as well as emotional conditions. He specifically points out that heart rate variability cannot be used if the subject is required to speak, since this affects the pattern of respiration and hence the heart rate. Nor can it be used if the test situation involves muscular activity, since this too affects heart rate. The same applies to the environmental temperature.

Assessment of muscular work loads

While oxygen uptake and heart rate may indeed be used to assess the severity of muscular work in general, these parameters are not always suitable for the monitoring of loads on local muscle groups (Malmquist *et al.*, 1981). Under certain conditions, however, as already pointed out, the recording of heart rate is quite a sensitive index of muscular engagement, and may even respond to minor changes in posture (Hanson and Jones, 1970).

On the other hand, electromyographic recordings may be used to measure the load on single muscles, or groups of muscles. This can be done by using surface electrodes. It is quite evident that the integrated electromyogram (EMG) does reflect the magnitude of the muscle engagement (see Figure 14.2, Chapter 14) and may be used to measure the exerted force in per cent of the maximal voluntary muscle strength. For a more detailed discussion of the principles underlying the use of electromyography in the evaluation of local muscular load, see Hagberg, (1981). According to Bobet and Norman (1982), the average electromyogram may also provide useful information in the investigation of whole body tasks. It should also be kept in mind that a straightforward electromyogram recorded with the aid of surface electrodes and a conventional electromyograph without integration may give a reasonable indication of the degree of muscle involvement in the case of simple tasks (see Figure 14.1, Chapter 14). It is also possible to combine the recording of EMG with portable, battery-operated loggers, such as the previously mentioned Squirrel meter/logger. For a further discussion of this subject, see Chapter 14.

Chapter 5
Methods for the assessment of the working environment

Assessment of thermal stress

The temperature of the environment is one of a variety of factors affecting human performance (see Figure 3.1, Chapter 3). The hot end of the scale is more problematic than the cold end, for we can protect ourselves more easily against overcooling than overheating, by adding protective clothing, etc.

The state of thermal comfort is generally characterized by core temperatures between 36.6 and 37.1 °C, and skin temperatures between 32.0 and 35.5 °C (Precht et al., 1973). At temperatures substantially higher than these optimal levels, both physical and mental performance may deteriorate due to a complicated interplay of physiological and pathophysiological processes. In excessive heat, a larger part of the available blood volume has to transport heat from the interior of the body to the skin where it can be dissipated in order to prevent overheating. This may limit the oxygen transporting capacity of the blood circulation, since phylogenetically temperature regulation appears to have priority over oxygen transport (Hardy, 1967). In addition, prolonged heat stress may lead to hypohydration, which in itself impairs performance, especially endurance. In addition, prolonged heat strain may impair mental and psychomotor functions, affecting performance. At temperatures substantially lower than the optimal levels, a lowering of the body temperature may partially be prevented by a reduction in the peripheral blood flow, causing a drop in the skin temperature in an effort to reduce heat loss. In extreme cases this may cause reduced neuromuscular function, including impaired manual dexterity (Wyon, 1982), partly due to reduced nervous conduction velocity (Vangaard, 1982). With reducing core temperature, physical performance is reduced by as much as 8 per cent per °C in maximal exercise of 3–8 min duration (Bergh, 1980). The impaired manual dexterity may be a contributing cause of industrial accidents, and prolonged severe cold exposure may lead to local cold injury and even hypothermia.

It is therefore of considerable practical importance to be able to assess the magnitude of the environmental thermal stress, and the worker's physiological reaction to it, in order to ensure optimal conditions for health and productivity.

Assessment of environmental heat stress

During the past several decades, considerable effort has been devoted to developing new methods for the assessment of environmental heat exposure at different places of work and refining some of the old heat stress indices, such as the Effective Temperature (ET), and the Wet Bulb-Globe Temperature (WBGT). By such indices one tries to predict the heat load on the individual. (For references, see Rodahl and Guthe, 1988.) These indices are usually based on data from physical measurements of air temperatures (dry bulb), radiant temperature (globe), relative humidity (wet bulb), and air velocity. A measurement of the individual's metabolic rate is also often included, i.e. the energy demand of the work is measured or estimated (Kerslake, 1972; I. Åstrand *et al.*, 1975). The WBGT index is rather complicated and the recording of it is time-consuming. In the case of a predominantly dry, radiant heat environment, the much simpler and faster reacting Wet Globe thermometer (WGT) or Botsball, devised by Botsford (1971), records similar values; the results are interchangeable with a simple formula (Ciriello and Snook, 1977, Beshir *et al.*, 1982).

A number of investigators have attempted to evaluate the different indices and compared them under a variety of conditions. Brief and Confer (1971) measured five heat stress indices simultaneously under varying environmental conditions in an environmental test room. They found that the WBGT and the WGT were best suited to hot work situations where radiant heat energy is present, and that the Effective Temperature (ET) was the preferred index for hot, moist conditions.

Botsball readings were compared with the WBGT at 30 different places of work in industrial plants (Beshir *et al.*, 1982). The measurements were taken twice daily over a one-year period for a total of 13,489 observations. The relationship in °C between the WGT and WBGT was found to be : WBGT = 1.01 WGT + 2.6, and the correlation coefficient was very high (R = 0.956). Beshir emphasizes the advantages with the Botsball thermometer in terms of convenient size, ruggedness and ease of use, and the fact that it may easily be located near the worker and thus permit an accurate evaluation of his working environment.

Ciriello and Snook (1977) investigated the relationship between the

Botsball temperature and WBGT under 210 different environmental conditions in an environmental chamber. The results indicate that as long as wind speed, humidity and radiant heat can be estimated fairly accurately, the WBGT can be predicted from WGT within ±0.4 °C. Only when adequate knowledge of the mentioned environmental conditions are not available, did the predictability fall to ±2.9 °C. It should be kept in mind that in the case of practical application in industry, a difference of some 3 °C may only be of academic significance.

A number of different heat stress indices were compared by Pulket *et al.* (1980), in a hot–humid environment. They concluded that, because of its ease of use, the WGT or the CET (Corrected Effective Temperature) index may be the best choice for preliminary industrial surveys in hot–humid environments.

The empirical relationships between several heat stress indices were examined by Mutchler and Vecchio (1977), including WBGT and WGT, derived from a series of studies in 14 representative, hot industries. Essentially they verified the findings of others that the differences between the indices are not dramatic. They point out that the predicted values for WBGT from selected values of WGT vary only slightly among investigators, the maximum difference being 2.2 °C.

On the basis of measurements of heat stress indices in 200 places of work, Locati *et al.* (1974) concluded that indices such as WGT and the WBGT could be correlated with sufficient accuracy.

Plunkett and Carter (1974) have described some of the problems encountered when trying to apply the WBGT index to assess heat stress in three Alcoa smelting plants. They showed that all work places investigated had two-hour, time weighted WBGT values above the recommended levels, with standard deviations of considerable magnitude, without this being associated with significant risks of heat disorders. They pointed out that Alcoa had about one reported heat illness per one million man-hours worked, and that many of the cases reported as heat illness were not truly cases of such disorders. Their investigated had two-hour, time-weighted WBGT values above the meaningful time-weighted average WBGT, when applied to aluminium smelting activities, where workers are exposed to conditions which vary from place to place and from day to day, makes this procedure impractical as an index to determine work practices. They were unable to determine any significant correlation that would allow them to use one, or a few, monitoring stations to represent an entire pot-room.

Malchaire (1976) has made a critical evaluation of the readings attained with different natural wet bulb and wet globe thermometers, with particular emphasis on design specifications. He concludes that there is a major drawback with the WBGT and WGT indices in that they are dependent on the design characteristics of the measuring instruments.

In 1969, Taylor *et al.* described an instrument for direct measurement of the Wet Bulb Globe Temperature Index, with a computer based on the mercury-in-glass thermometer. More recently, the Brüel and Kiaer Company has produced a commercially available WBGT recorder. Rapp and Aubertin (1985) have also developed a computer program for the calculation of mean radiation temperature in work places. Mutchler *et al.* (1976) have even developed a model for the prediction of WBGT in a hot environment, based on outside weather conditions.

Kamon and Ryan (1981) have suggested an 'Effective Heat Strain Index' (EHSI) using a hand calculator with memory for on-the-site evaluation of prevailing ambient conditions. The input to the programmed calculator includes dry-bulb, wet-bulb and globe temperatures, an estimate of metabolic rate and air movements. The index is based on a program for the calculation of the total heat balance and the efficiency of sweating.

Ramsey and Chai (1983) have examined the inherent variability in heat-stress decision rules. They conclude that rather than being lulled into a sense of accuracy which may result from obtaining highly accurate thermal measurements, it might be more appropriate to utilize such measurements as general indicators of thermal stress, and then to provide reasonable work practices for reducing the risk of heat disorders. They suggest that the use of a simplified set of decision rules can basically serve the same purpose as a seemingly more precise approach.

Wet bulb-globe temperature (WBGT)

The WBGT index was introduced in 1957 (Yaglou and Minard, 1957) as an improvement of the ET (Effective Temperature), taking radiation into account. The index is based on a simple formula, which has been the subject of several modifications over the years. This may be confusing since it reflects different parameters and different weighings of some of the parameters (Botsford, 1971). Nonetheless, it is a meaningful manner of describing the total heat stress of the environment, under specific circumstances. A major disadvantage with the WBGT is the time it takes for the instrument to equilibrate (up to 30 min).

Based on the WBGT, the American Conference of Governmental Industrial Hygienists (ACGIH, 1976) have suggested tolerance limits for heat stress exposure for industrial workers. This includes continuous work and intermittent work (75, 50 and 25 per cent work; 25, 50 and 75 per cent rest), and work load (light and moderately heavy). The recommendations are based on the assumption that the rectal temperature of the worker should not be allowed to exceed 38 °C.

There are a number of obvious objections to these recommendations. In the first place they are based too much on theory and assumption rather than on physiological observations at actual work places. The basic assumption of a rectal temperature ceiling of 38 °C is unfounded in the first place. The values stipulated for intermittent work are inadequately based on actual observations. Furthermore, industrial work is in reality seldom performed in this manner or according to the proposed schedule between rest and work. Finally, there is no direct evidence showing exact environmental limits affecting health. This depends as much on how work is done, the length of each exposure and cooling–off period, protection, fluid intake, etc. And in the final analysis the exposed worker himself is usually able to tell how he feels, and how hot it is, more so than any index is able to do.

Parikh *et al.* (1976) undertook a field evaluation of the WBGT index under the conditions of severe heat stress experienced by workers in Indian industries. They recorded heart rate and environmental temperature (WBGT) in exposed workers in 9 small foundries in three seasons. They found that while the WBGT far exceeded the WHO proposed limits, the average heart rate during the shift rarely exceeded 110 beats per min.

Ljungberg *et al.* (1979) studied the effects of sedentary work in two different climates: 40 °C and 40 per cent relative humidity; and 32 °C and 80 per cent relative humidity. They concluded that the common indices such as WBGT did not discriminate between the two climatic conditions while indices based on heat balance analysis or calculations of sweat rate could more accurately predict the physiological and psychological strain.

The wet bulb temperature (WBT)

The WBT, based on the Botsball thermometer, introduced by Botsford (1971), is an instrument which combines air temperature, humidity, wind and thermal radiation into a single reading, expressing the thermal stress of the environment. The instrument is quite simple and easily adapted to industrial use. It consists of a small 60 mm hollow black globe, covered with a double layer of black cloth, which is continuously moistened by water seeping from a reservoir tube attached to the globe. The stem of a dial thermometer passes through a plastic tube along the center-line of the water reservoir and into the globe, thus sensing its temperature.

When placed in a hot area, the globe is heated by the surrounding air and by radiant heat from any hot surfaces in the surroundings. It is cooled by the evaporation of water from the globe surface, depending on air humidity and wind or air movement. The wet globe reaches an equilibrium temperature when these heating and cooling effects come into balance, which usually takes about five minutes. The 'Botsball' temperature is read directly on the dial thermometer.

The Botsball thermometer is light in weight and may be mounted on a simple stand or hung in a suitable location representative of the place of work. Vibrations transmitted through the support may cause an excessive amount of water to escape from the reservoir. This problem is avoided by hanging the globe in elastic bands. Errors in the thermometer readings are less than 0.5 °C, but the thermometer should be calibrated regularly to maintain this accuracy. If the Botsball thermometer is to be used in a strong magnetic field, an error of up to 1.7 °C may be expected. A shielded Botsball thermometer is to be preferred in such locations.

This Botsball temperature has been found to correlate quite well with other comparable indices. In our experience the WGT readings are in reasonably good agreement with the WBGT under conditions of dry, radiant heat, and since the former is easier to use and reacts faster, we consider it preferable for practical field applications.

The continuous recording of the environmental temperature indices

The previously mentioned Squirrel logger, produced by Grant Instruments (Cambridge) Ltd., is ideally suited for the continuous recording of both the WBGT and the WGT index (Figure 4.12, Chapter 4). It is simple to operate, dependable and inexpensive. The data accumulated in the logger's internal memory can easily be transferred in the field to a miniature, transportable computer, fitted into a suitcase, or to a regular IBM compatible personal computer in the office, for further statistical treatment of the data, for print out or for the making of graphs.

The temperature probes for the Squirrel have been compared with a standard mercury thermometer and found to be quite accurate. The WBGT has three probes, occupying three channels in the Squirrel (1 = dry bulb temperature; 2 = globe temperature; 3 = wet bulb temperature) (Figure 5.1). The WBGT index may be calculated by the following formula:

WBGT = (Channel 1 × 0.1) + (Channel 2 × 0.2) + (Channel 3 × 0.7)

The advantage with the WGT or Botsball thermometer is that the WGT index can be read directly, and that it occupies only one channel in the Squirrel logger (Figure 5.2). The reading of the Botsball (WGT) can be converted to WBGT by the following formula:

WBGT = (WGT × 0.0212) + (WGT × 0.192) + 9.5

The converted readings of the WGT were compared with the WBGT

Figure 5.1. Photograph showing the WBGT-assembly for environmental heat stress recording, plugged into the Squirrel meter/logger.

Figure 5.2. The Botsball thermometer (WGT) for the assessment of environmental heat stress, mounted on the worker's helmet.

readings, the two instruments being placed side-by-side in different places in and around the laboratory. The mean difference between the two sets of readings was 0.5 °C, which is quite satisfactory. A similar comparison between the WBGT and WGT converted into WBGT was made in a ferroalloy plant, Salten Verk, in the north of Norway in the month of July. Here, under actual field conditions with changing radiant heat and considerable air movement, the mean difference was 1.5 °C, the difference being less the higher the temperature (range of difference 0.2–2.9 °C). For all practical purposes this degree of difference is quite acceptable in view of the fact that the environmental temperature in the plant may fluctuate by as much as 10 °C from hour to hour in the same place.

Another advantage with the Botsball thermometer is that it can be mounted on the helmet and thus carried by the person, allowing the environmental temperature to be recorded where the worker is actually located at any time (see Figure 5.2). The only drawback with this arrangement is that walking briskly causes the WGT to drop a couple of degrees, due to air motion causing increased evaporation from the globe surface, hence cooling the probe.

The ideal solution is probably to use the Botsball thermometer attached to the worker initially for the purpose of surveying the temperature distribution in the working area, and then to place the Botsball thermometer stationary at the location where the temperature is most extreme, or most representative. The placing of the temperature probe inside the Botsball thermometer is not critical, providing the sensor is located within the globe of the thermometer.

Assessment of heat strain

It is thus clear that an assessment of the environmental heat stress can be made with the aid of the Squirrel logger using a WBGT assembly or a Botsball WGT instrument. However, the environment is one thing, but the human reaction to that environment is something else, and of far greater importance. For this reason, greater emphasis should be placed on the assessment of human response to the heat stress encountered by workers during the performance of their everyday work at their actual work places. Without such information any discussion of safe or upper limits of exposure would seem meaningless.

For the assessment of the physiological reaction to heat loads imposed by the environment, the body heat content (S) is a most meaningful index of body heat gain or body heat loss. It can be calculated by the formula:

$$S(\text{in kcal}) = 0.83 \times W \text{ (body weight in kg)} \times (\text{rectal}$$
$$\text{temperature} \times 0.65 + \text{mean skin temperature} \times 0.35)$$

Rectal temperature

Rectal temperature mirrors core temperature and may by itself reflect body heat gain or body heat loss. Under such conditions rectal temperature may be used as an index of heat stress, as long as sufficient consideration is given to the fact that it is a slow reacting parameter. In most cases of industrial heat exposure it takes some 45 min for the rectal temperature to reach a plateau. This emphasizes the importance of continuous recording of the rectal temperature by portable electronic loggers such as the Squirrel. This, however, requires co-operative subjects, and patience and persuasion on the part of the investigator. However, when correctly carried out, the continuous recording of the rectal temperature in workers in the course of their normal work and leisure is indeed a most revealing parameter in terms of thermal strain (see Figure 5.3).

Ideally, the temperature sensor should be placed no less than 8 cm into the rectum in order to achieve stable readings representing deep body temperature. This may conveniently be done by supplying the rectal temperature sensor with a taped knob 8 cm from the end and to

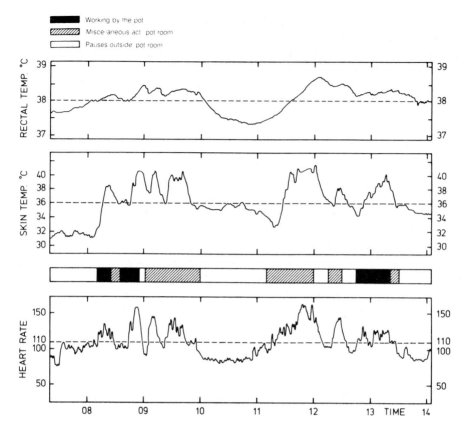

Figure 5.3. Rectal and skin temperature and heart rate in a pot tender, working in an aluminium plant.

insert the sensor far enough into the rectum for the knob to be placed inside the rectal sphincter, and to tape the wire to the subject's lower back to prevent it from being pulled out. For hygienic reasons, the sensor may be placed inside a thin disposable plastic envelope before insertion.

For practical reasons, it might, in some cases, be simpler to record the temperature inside the ear as an index of deep body temperature, replacing rectal temperature (T_r) in the body heat content equation. For a comprehensive discussion of the different deep body or core temperatures the reader is referred to Chapter 13 in Åstrand and Rodahl (1986). For basic research purposes it has long been customary to measure the tympanic temperature as an expression of the deep body temperature, close to the location of the thermoregulatory center in the brain, by placing a thin thermocouple into the ear, touching the lower part of the ear drum (Benzinger and Taylor, 1963). In the case of field measurements in working individuals at their industrial work places, this may not be any easier to do than using a rectal thermocouple. The manipulation of the thermocouple into the ear duct and placing it in touch with the ear drum requires some professional skill. In addition, it may be painful to the subject, even when using a very thin and flexible thermocouple wire. In any case, it is necessary to plug the external opening of the ear duct to prevent the ambient air interfering with the readings. Even so it appears possible that in very hot environments the tympanic temperature may, to some extent, be affected by the hot ambient air. For a further discussion of this subject the reader is referred to Guidry and McDowell (1966), Wurster (1968), Moore and Newbower (1978), Palca (1981), Uchino *et al.* (1986), Cabanac *et al.* (1987), and Brinnel *et al.* (1987).

For the purpose of applied studies or less sophisticated research projects a simple measurement of the temperature inside the outer ear duct may be used, providing the ear duct is plugged and the ear is properly covered during the measurements, and providing sufficient time (no less than 20 min) is allowed for the ear duct temperature to equilibrate with the surrounding tissue temperature. If so, the ear duct temperature, in our experience, is about 0.6 °C lower than the rectal temperature, changing synchronously with the latter. Suitable ear plugs for this purpose compatible with the Squirrel logger are commercially available.

In a field study at a ferroalloy plant in Norway (Salten Verk), a reasonably good correlation was observed between rectal temperature and ear duct temperature measured in this manner, in the same individual, exposed to a considerable heat stress. The mean difference was 0.6 °C.

For a rough estimation of deep body temperature, the much simpler oral (sublingual) temperature may be used, providing care is taken to keep the sensor or thermocouple in place under the tongue, and providing the subject keeps his mouth completely closed. Oral temperatures recorded in this manner are in most cases about 0.5 °C lower than the rectal temperature.

Skin temperature

Skin temperature can be measured relatively easily, either by thermocouples, thermistors or disposable probes. It may be used as an index of local heat exposure, but cannot by itself be used as an indication of general heat stress or thermal balance. In our experience, however, it can indeed be used as an indication of how close the worker is to the heat source (see Figure 5.4, which shows the skin temperature in a

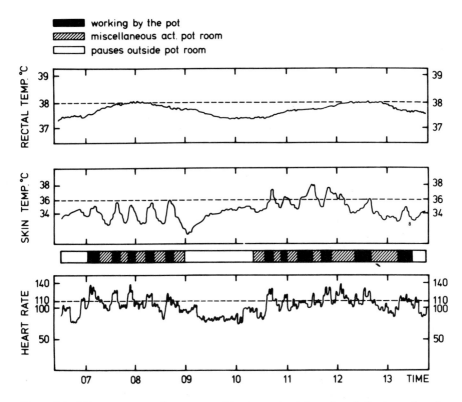

working by the pot
miscellaneous act. pot room
pauses outside pot room

Figure 5.4. *Skin temperature of a tapper working in an aluminium plant, indicating radiant heat exposure each time he is close to the pot, tapping the finished metal.*

tapper in an aluminium plant rising abruptly each time the tapper approaches the pot). This illustrates the importance of proper shielding of the temperature sensor under conditions of intense radiating heat in order to obtain real values of skin temperature. Otherwise, it should be noted that the skin temperature recorded on the inside of the thigh to some extent resembles mean skin temperature under most normal circumstances (Ramanathan, 1964) (Figure 5.5).

Skin temperature is the best single temperature index of warm and cold sensation, both in warm and cold environments. Skin wettedness is the best single index of warm discomfort (Gagge *et al.*, 1974).

For a comprehensive review of the practical assessment of heat stress, see Brotherhood (1987).

In summary, the miniature electronic Squirrel logger, which can easily be carried on the individual, is ideally suited for the continuous logging of both the environmental heat stress, in terms of the WGT or the WBGT and the resultant heat strain in terms of changes in body heat content, reflected by changes in core and shell temperatures.

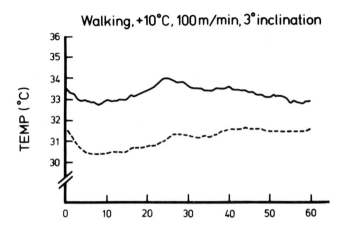

Figure 5.5. Mean skin temperature (solid line) and skin temperature on the inside of the thigh (dotted line) in a subject walking on a treadmill at 10 °C, lightly dressed, at a speed of 100 m·min⁻¹, inclination 3°.

Furthermore, all these parameters may be recorded simultaneously by the same Squirrel logger. With the new series of multichannel Squirrel loggers a number of other pertinent parameters such as heart rate (as an indication of work stress), muscular tension, CO-concentration in the ambient air etc., can be continuously recorded simultaneously on the same logger.

Assessment of the insulative value of protective clothing

Most types of clothing provide thermal insulation and will protect the individual not only from cold by preventing heat loss, but also to some extent from heat by providing shielding from radiant heat gain, and by absorbing heat from the environment.

The insulating value of clothing is usually expressed in Clo units. The Clo unit is a measure of the resistance of clothing to sensible heat transfer (Gagge *et al.*, 1941). A Clo unit is the thermal insulation which will maintain a resting individual indefinitely comfortable in an environment of 21 °C, relative humidity less than 50 per cent, and air movement 6 m per minute. One Clo is equal to $0.155 \cdot m^2 \cdot °C \cdot W^{-1}$, where m^2 = surface area in m^2, °C = temperature difference between skin and air, and W = watt, i.e. energy output. Practically speaking, one Clo is roughly the amount of insulation provided by an ordinary suit of clothing considered adequate to keep the wearer in thermal balance during regular indoor use. As the insulating value is primarily a function of the amount of trapped air in the clothing, the type of fiber used is of less importance than how the fibers are used in terms of providing air space between the fibers.

Since the pioneering efforts of Gagge, Burton and Bazett in 1941, the US Quartermaster Laboratories at NATIC, Mass. have made significant contributions to the development of concepts and methods for the quantitative analysis of heat transfer through clothing. More recently, Holmér (1983) has supplemented this effort with particular emphasis on how to evaluate objectively the protection offered by a clothing system against the thermal stress of the environment and how to compare differences between garments.

The individual's heat exchange with the environment may be expressed in terms of the standard heat exchange equation:

$$S = M \pm R \pm C - E$$

where S is stored body heat; M is the heat produced metabolically; R is the radiant heat exchange; C is convective and conductive heat exchange; and E is evaporative heat loss.

S can be calculated from the measurement of core and mean skin temperature, as previously described, according to the formula:

$$S = 0.83 \cdot W \, (0.65 \cdot T_r + 0.35 \cdot T_s)$$

where W is body weight in kg; 0.83 is the specific heat of the body; 0.65 and 0.35 are the fractions assigned to the core (rectal) and mean skin temperatures respectively (Burton, 1935).

M, *the metabolic heat production*, can be determined by measuring the individual's oxygen uptake, or assessed indirectly by recording the individual's heart rate and converting it into energy expenditure by methods previously described. The uptake of one liter of oxygen equals a heat production equivalent to 4.8 kcal (20 kJ).

R, *the radiant heat exchange*, is a function of the temperature difference between the individual and his surroundings. This may be assessed by a globe thermometer, i.e. a mercury thermometer, the bulb of which is placed inside a hollow spheric black copper globe with a standard diameter of 15 cm. For the purpose of industrial surveys, where the radiation intensity may be rapidly changing and extremely variable, satisfactory results may also be obtained by using a smaller black globe, connected with a meter/logger such as the Squirrel.

C, *convective heat loss or gain*, is affected by the temperature of the air, which can be measured by an ordinary mercury thermometer or by a temperature sensor (thermocouple) connected to a meter or a logger, such as the previously described Squirrel. The temperature sensor should be properly shielded from thermal radiation; for example, by the use of a piece of tinfoil placed in such a manner that free air passage is provided around the thermometer, thermocouple or thermistor.

E, *evaporative heat loss*, is heat lost from the body due to the evaporation of sweat. This amounts to 580 kcal (2.4 MJ) per liter of sweat evaporated. Sweat loss may be estimated by recording nude body weight on a scale, before and after the observation period, preferably with an accuracy better than ± 20 g. Food and fluid ingested, as well as stools and urine voided, during the period of observation, have to be weighed and included in the calculation. The recorded weight loss also includes weight loss due to respiratory gas exchange (Snellen, 1966) which may be accounted for by using the formula:

$$Cge = VO_2 \,(1.977 \cdot R - 1.429)$$

where

Cge = weight loss in g per min due to respiratory gas exchange.
VO_2 = oxygen uptake in liters per min.
R = respiratory quotient (CO_2/O_2)
The figure 1.977 is the weight in g of 1 liter CO_2
1.429 is the weight in g of 1 liter O_2

The evaporative loss (E) is affected by the *humidity of the air*, in that it affects the rate of sweat evaporation. The humidity of the air can be determined with the aid of a sling psychrometer or an electronic humidity sensor. The *air movement* affects both evaporative heat loss (E) and convective heat exchange (C). It is usually measured by a hot wire anemometer.

S, *the amount of stored heat in the human body*, amounts to about 2000 kcal for a 70 kg individual in thermal neutrality ($T_r = 37\,°C$, $T_s = 33\,°C$). An increase in mean body temperature of about 1 °C per hour is equivalent to a rate of body heat storage of some 40 watts per m^2.

Heat transmitted via the skin surface to the environment (via R and C) amounts to some 80–90 per cent of the total heat loss. This heat has to pass through all the layers of clothing, covering up to 95 per cent of the body surface. This transfer takes place both as dry or 'sensible' heat transfer, and as humid or 'latent' heat transfer. The dry heat transfer (H_{dry}) accounts for the heat transported by convection, conduction, and radiation, and is defined as:

$$H_{dry} = \frac{T_s - T_a}{R_c}$$

where T_s = mean skin temperature, T_a = ambient air temperature,

R_c = resistance to dry heat transfer of clothing and air layers, expressed in Clo units. The humid heat loss (H_{hum}) accounts for the latent heat of the evaporated sweat at the surface of the skin and transported as vapour through the clothing and air layers. This humid heat loss can be determined according to the formula:

$$\frac{P_s - P_a}{R_c}$$

where P_s = mean water vapour pressure at the skin surface; P_a = ambient water vapour pressure (expressed in Pascal); R_c = the resistance to evaporative heat transfer by clothing and air layers (Holmér and Elnäs, 1981).

Holmér and Elnäs (1981) have developed a laboratory method to determine the effective evaporative resistance of clothing *in vivo*, based on direct measurements of the water vapour pressure gradient between skin and ambient air and the steady state rate of evaporative heat loss. Air is sampled by tubes from six different places on the skin under the clothing and pumped to an oxygen analyzer via a mixing chamber. Water vapour pressure is derived from measurements of oxygen partial pressure in the atmospheric air, using the general gas law. Evaporative heat loss is determined by continuous weighing of the subject on an electronic balance. Sophisticated methods such as this require elaborate laboratory or climatic chamber facilities, where certain aspects of real life can be simulated, but not entirely duplicated. There is therefore a need for simple, yet basic methods for field use to evaluate objectively different physiological aspects of practical clothing worn under real-life conditions. One very simple possibility of assessing the permeability of different types of clothing for water vapour is to determine the amount of sweat loss by weighing the subject before and after the observation period, and by weighing the clothing before and after in order to determine how much of the produced moisture has accumulated in the clothing. The difference is a rough estimate of the amount of moisture which has escaped through the clothing, and may be taken as an indication of the moisture permeability of the clothing.

The thermal properties of clothing systems as such are routinely determined, either by the guarded hot plate technique, the thermal manikin ('copper man') or by means of direct or partitional calorimetry. A rough physiological comparison of the insulating value of specific clothing assemblies may also be attained simply by measuring heat production in terms of oxygen uptake, and heat loss as evidenced by changes in stored body heat (S), using the same subject as his own control under standard climatic chamber conditions. An example of such an evaluation is presented in Figure 5.6, which represents the mean of five subjects. In this study similar garments made of nylon pile and wool pile were compared in paired experiments at rest for 1 hour

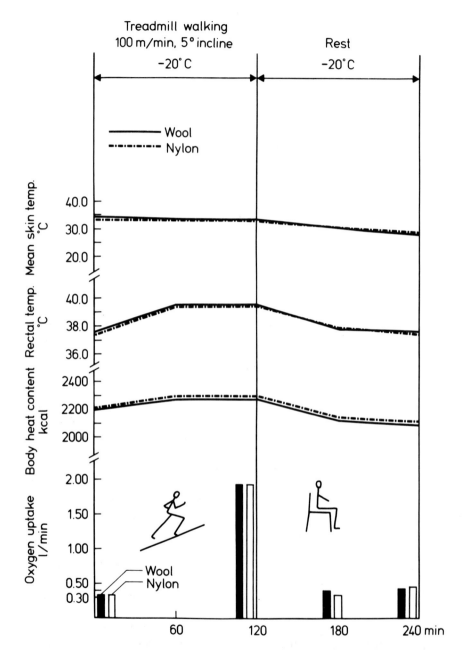

Figure 5.6. Assessment of thermal insulating value of identical garment made of nylon pile and wool pile, by measuring heat production (oxygen uptake) and heat loss, assessed by changes in body temperature (rectal and skin) and body heat content.

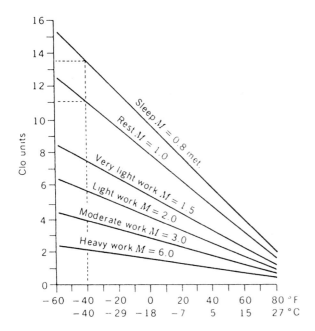

Figure 5.7. Insulating requirement (Clo) in the cold when protected from the wind during different rates of heat production, at different environmental temperatures. One 'met' is the metabolic rate (M) of a resting man (3.5 ml $O_2 \cdot kg^{-1}$) (from Burton and Edholm, 1969).

followed by 2 hours of treadmill walking at 100 m per min and at a 5° incline, followed by a 2-hour rest at $-20\,°C$ in a climatic chamber. The accumulation of moisture in the experimental clothing was assessed by weighing the garments before and after the experiment on a scale with an accuracy of ± 10 g. In this study, no significant difference could be detected between the two types of garments in terms of thermal insulation, nor in the ability of the two types of fabric to allow free escape of moisture produced by sweating during the physical activity (Rodahl *et al.*, 1974a).

The amount of insulation required in the form of clothing in the cold, when protected from the wind, during different degrees of physical activity and hence different rates of heat production, is illustrated in Figure 5.7. From this figure it appears that in order to remain in thermal balance, a person sleeping at an ambient temperature of $-40\,°C$ needs thermal protection in the form of clothing or a sleeping bag equivalent to about 12 Clo units, which can best be provided in the form of furs (Figure 5.8). However, when moderately active, only the equivalent of 4 Clo units may be needed because the person's body heat production may now be as much as three times greater than it was when sleeping

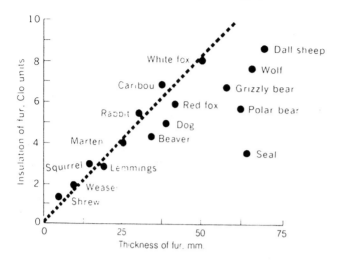

*Figure 5.8. Insulating value (in Clo units) of different furs
(redrawn from Scholander et al., 1950).*

because of the increased metabolic rate due to the increased physical
activity. This insulating requirement is provided by the ordinary winter
uniform worn by our soldiers in the Arctic.

   The problem of clothing in the cold when heavy physical activity is
alternated with rest periods may be complicated, since there is no single
item of clothing capable of both protecting against cold at rest and of
facilitating heat dissipation during heavy work. The best one can do is
to unbutton the outer garment during work and to button it up again
during inactivity.

   The surface of the hands represents about 5 per cent of the total
surface area of the body. In a clothed individual, up to 20 per cent of the
heat produced by the body may be eliminated through the hands (Day,
1949). Because of the remarkable ability of the hands to vary the degree
of their blood perfusion, they may play a significant role in the
temperature regulation of the body (for further discussion, see Åstrand
and Rodahl, 1986).

   Olesen *et al.* (1982), observed that clothing insulation, determined by
a manikin, is influenced by posture, being some 8–18 per cent less in
sitting than in standing posture. A further 10 per cent decrease was
observed when the manikin was moving (cycling), due to the pumping
of the air in and between the clothing layers, caused by the body
movement. The need to take this pumping effect into account when
considering the effective insulation of clothing worn by living subjects
has been emphasized by Vogt *et al.* (1983) who have shown that the
pumping effect can both increase and decrease the resultant clothing
insulation.

Assessment of the lighting conditions at the individual place of work

Although there are considerable individual differences in the subjective judgment of what may be considered adequate lighting conditions at a specific place of work, older individuals generally require much more light than do younger individuals in order to see what they are doing.

The problem of lighting may be a matter of insufficient light falling on the observed object (i.e. illumination) or it may be a matter of glare. Students of lighting distinguish between *illuminance*, that is the amount of light falling on the object one is looking at, and *luminance*, which is the amount of light reflected from it. The former is measured with an illumination meter which indicates the amount of light falling on the surface on which it is laid. The reading is expressed in lux or foot-candle units. The luminance is measured by a photometer. It is positioned at the location of the operators head, so as to measure the amount of light reflected towards his eyes. The readings are expressed in candelas per square meter, or in footlamberts. Detailed information for the use of these instruments can usually be found in the instruction manual supplied by the producers of the instruments. For more detailed information on the lighting measurements, the reader is referred to Kaufmann and Christensen (1972), and Rodgers and Eggleton (1983).

Assessment of noise levels

In most cases, a standard sound-level meter can be used for routine evaluation of noise. It can be used to identify the sound pressure level in decibels on the A scale, i.e. dBA. It can also be used to assess the potential harmful effect of the noise in terms of hearing loss and communication interference.

When there is a need to characterize the noise according to frequency bands, precision sound meters, equipped with filters, are used. When the sound level is measured in each band, the relative proportion of high, medium, and low frequencies can be determined. This may be useful in order to determine the effectiveness of various noise-control measures, and to identify specific parts of noise-producing equipment which need to be sound-shielded.

For the recording of the individual worker's noise exposure during the entire work shift or work day, portable noise dosimeters, carried on the person, are available, such as the battery-operated, microprocessor-based M-28 Noise Logging Dosimeter produced by Quest Electronics, which offers direct reading display and output to printers and computers. For more practical guidelines for the use of noise measurement equipment, see Veit (1971), the ISO R 1999 publication (1971), Metrosonics (1981), and Rodgers and Eggleton (1983).

Assessment of local and whole-body vibration

Prolonged use of vibrating hand tools may lead to the classical hand–arm vibration syndrome. This includes the Raynaud's phenomenon, characterized by cold-induced white fingers. Our present knowledge of this disabling condition is still incomplete, especially knowledge of the quantitative relationship between exposure dose and effects (Gemne, 1987). Above all there is a need for a battery of well-documented methods for making valid diagnoses which may enjoy general acceptance. This is especially important in view of the possible legal consequences of this diagnosis in insurance compensation cases.

Vibration itself may tend to increase the force exerted to hold the vibrating tool, and thus enhance the danger of vibration injury. This close relationship between force and vibration, and the difficulty of measuring the precise magnitude of the force exerted, and the exact level of the vibration intensity in manual work, makes it difficult to determine the relationship between dose and effect (Armstrong *et al.*, 1987).

In the diagnosis of suspected vibration syndrome, the case history is of major importance (Ekenvall, 1987). Since vibration affects not only the musculoskeletal systems, but also the peripheral circulatory and nervous systems, these manifestations should be included in the test battery used to diagnose the hand–arm vibration syndrome. Matoba and Sakurai (1987) have suggested the following battery of tests: peripheral circulatory function tests, which include skin temperature measured by thermistors under specific temperature conditions; the nail compression test; and the cold provocation test involving a 10-min immersion of the extremity in cold water at 5 or 10 °C. The peripheral nervous function test includes testing grip strength, pinch strength and tapping speed.

Saito (1987) has shown that strict restrictions on the operating time of vibrating hand tools, and on the age of the workers using them, can largely prevent the vibration syndrome.

Whole body vibration is even more complicated to assess than is local hand–arm vibration. One of the most apparent effects of whole body vibration is fatigue, a phenomenon which is difficult to assess objectively, but easily felt subjectively. A fact which anyone can verify who has experienced prolonged flights in oldfashioned shaky military aircraft, or even more so in modern helicopters.

The vibration or shaking can be characterized by the two parameters: the speed of the movement (acceleration) and the frequency by which the movement changes back and forth. The frequency is measured in Hertz (Hz). The acceleration is measured in meters per second squared $(m \cdot s^2)$. The acceleration is also referred to as amplitude, and measured in dB.

The vibration effect on the human body is most pronounced in the low frequency range of 1–100 Hz. The human body has the lowest tolerance to vibration in the 4–8 Hz range. This is due to the amplification of the vibration by the natural resonance of the human body. Generally speaking, frequencies up to 10 Hz have the most detrimental organic effects. Guidelines for the measurement and evaluation of human exposure to vibration are given in the ISO publications 2631 (1974) and 5349 (1979).

Ambulatory assessment of blood pressure, muscle tension and heart rate

Portable devices are also available for ambulatory 24-hour intra-arterial blood pressure recordings (Tochikubo *et al.*, 1986), as well as non-invasive, portable blood pressure recording systems (Wolthuis *et al.*, 1981). A fully automatic, portable recorder of indirectly measured blood pressure in subjects awake and asleep, at home and at work, has been described by Schneider (1981). It records blood pressure every 15 min for as long as 10 hours or more. The validity of this device has been tested by comparing the results from this device with the standard method, and by comparing 45 determinations by this method in nine subjects with intravascular values.

A portable device for long-term surface EMG recording, storage, processing and display, has been developed and tested, both in the laboratory and in the field (Laurig and Smolka, 1980).

During recent years, a variety of portable, miniature, solid state, battery-operated heart rate recorders have been described (Crane, 1978; Hoodless *et al.*, 1980), in addition to the well-known Medilog and Vitalog recorders, already mentioned. The main advantage with the new generation of Squirrel meter/loggers (the 1200 series) is the possibility of recording a variety of interrelated parameters simultaneously by one and the same logger, and the immediate display of results. The use of such portable electronic loggers in physiology is largely the result of the technology of the space age and the introduction of the microprocessor (Hilfert *et al.*, 1980; Rodin *et al.*, 1984).

Assessment of air pollution

Assessment of toxic gases in the ambient air

Gases such as CO, SO_2, H_2S etc., represent an important aspect of air pollution as a consequence of modern industry, automobile traffic, power plants, etc. So far, this has primarily been an area of responsibility allotted to the industrial hygienists. But since the uptake of these gases as well as respirable dust, increases with increasing pulmonary ventilation as the result of increasing work rate, the problem of the pulmonary uptake of these products is also a matter of interest for the work physiologist, who has a special competence in the field of pulmonary physiology. We may therefore expect a much closer collaboration between the occupational hygienist, the work physiologist and the plant physician in the solution of occupational health problems than has been the case in the past.

In the case of carbon monoxide, well established, stationary systems for the monitoring and control of the atmosphere in different places of work have been used to measure the CO content of the ambient air. The disadvantage with these permanently installed systems is that they do not always necessarily measure and record the CO-concentration where the operators work. The portable, battery-operated CO-meters, on the other hand, may measure the CO-concentration at the location where the operators are working, but the concentrations may fluctuate so rapidly that it may be difficult to decide which reading to use. The advantage with the combination of CO-sensors and electronic meter/loggers capable of storing the recordings in their memories is that it provides for a more dynamic demonstration of CO-exposures from one moment to the next. An example of this is given in Figure 15.5, Chapter 15. But even so, the CO-concentration in the ambient air is one thing, the amount of this CO taken up by the blood is something else, and of more vital importance to the subject involved. Most of the existing tables showing the relationship between CO-concentration in the inhalable air and the amount of carboxyhemoglobin (COHb) in the blood at different lengths of exposure, refer to individuals at rest. In view of the fact that pulmonary ventilation may increase by a factor of 10 or more when changing from rest to work, the uptake of CO by the blood may be similarly increased. For this reason, it is necessary to take the level of pulmonary ventilation into account when considering CO-uptake in an individual working in a CO-containing atmosphere. This can be done by using a multi-channel logger, such as the previously described Squirrel, to record the CO-concentration of the ambient air. At the same time it can record heart rate, as an indication of work rate, and hence pulmonary ventilation. Or it can record pulmonary ventilation directly with the aid of an appropriate air flow sensor.

Finally, the COHb percentage saturation can be determined by a Squirrel-connected Bedfont CO-monitor, and thus the relationship between CO-concentration in the respirable air, and CO-uptake in the blood, at different work rates and different lengths of exposure, can be established. This is based on the established method of assessing COHb levels by measuring the CO-concentration in the expired air at the end of expiration, after breath holding, as used in the Bedfont EC 50 analyzer produced by Bedfont Technical Instruments Ltd, Sittingbourne, Kent, England (Jarvis *et al.*, 1986; Irving *et al.*, 1988). For further details see Chapter 15 where the merits of three commercially available CO-sensors are described. They are: The SABRE CO-analyzer, produced by Sabre Gas Detection Ltd, Guildford, Surrey, England; the DRÄGER COMOPAC CO-analyzer, produced by Drägerwerk, Lybeck, Germany; and the COMPUR MONITOX CO-analyzer, produced by Compur Electronic, München, Germany. These electrochemical sensors contain a working electrode, a counter electrode and a reference electrode in an acid electrolyte. The CO in the air sample is oxidized at the anode, and the electrical current thus produced is proportional to the CO-concentration.

Calibration with known CO-test gases in air showed a straight line correlation between CO-concentration and millivolt output in all three sensors (see Figure 15.2, Chapter 15 as an example). Recalibration tests several weeks later showed a high degree of stability.

All three sensors were used to measure the CO-concentration of the air at different locations in a silicon carbide (SiC) production plant. They were plugged into the same Squirrel meter/logger, so that the readings from the three sensors could be compared simultaneously with the readings of a conventional Commonwarn CO-meter. This comparison turned out to be rather difficult, due to the extremely rapid oscillations in the readings of the Dräger Commonwarn CO-meter, which was a direct reading instrument without memory for data storage, as opposed to the much slower reacting test sensors. Representative samples of ambient air from the production hall of the plant were therefore collected in large plastic bags, and the CO-concentration measured after thorough mixing, by the three test sensors, compared with the readings of the Commonwarn CO-meter, and in one case compared with the readings of the Ecolyzer CO-meter. The results showed a reasonable degree of agreement. At the 130 ppm-level, the readings with the Compur and the Commonwarn sensors were identical. The Dräger CO-sensor showed 6 ppm higher values and the Sabre 15 ppm higher than the Commonwarn reference values. At the 40 ppm level, the difference between the standard reference sensor and the three trial ones varied from 3 to 7 ppm.

It is thus evident that the CO-sensors that are already available, combined with a meter/logger such as the Squirrel, can for all practical purposes be used for the routine recording of CO-concentrations in the ambient air in a variety of industrial places of work. They can also be used for the purpose of assessing CO-concentrations in the ambient air in areas with heavy automobile traffic, in tunnels, etc.

Sensors compatible with loggers such as the Squirrel are also available for the continuous and integrated recording of a variety of other gases, such as O_2, CO_2, SO_2, and H_2S etc. combined with sensors capable of recording a variety of physiological reactions to environmental stress (temperature, heart rate, muscle tension, etc.). This offers a new opportunity of recording simultaneously by one and the same instru-

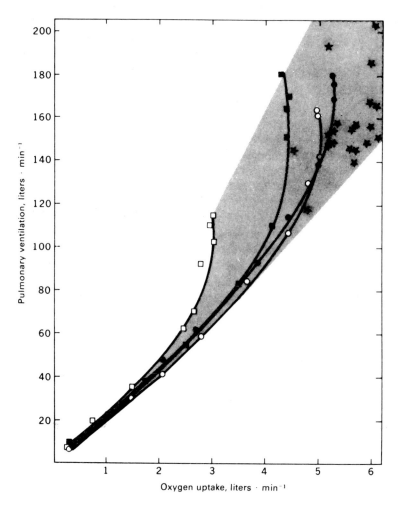

Figure 5.9. Relationship between pulmonary ventilation (l·min⁻¹) and oxygen uptake (l·min⁻¹) (redrawn from Saltin and Åstrand, 1967).

ment a number of interrelated parameters representing aspects of the working environment, and the worker's physiological reactions to it, while doing his job at his actual place of work. A major advantage with this system is that it is rugged, simple to operate and dependable. This renders it ideally suited for use by industrial personnel who may not be scientifically trained or experienced in sophisticated research techniques.

Assessment of respirable dust

Dust particles of a size which allows them to enter the respiratory tract are a major potential health hazard in modern industry. When the organism is exposed to respirable dust the amount inhaled and transported down the respiratory tract increases with the respiratory rate, which is a function of work rate (Figure 5.9). Thus, the heavier one works the greater the respiratory rate, and the more one inhales of the respirable dust particles suspended in the air one is breathing. For this reason, the industrial surveyor concerned with the worker's health hazards is not only concerned with the amount of respirable dust contained in the ambient air, but equally concerned with the worker's physical work load, and hence the magnitude of his pulmonary ventilation. With the modern electronic multi-channel loggers presently available, such as the Squirrel, it is now possible to assess the size and concentration of dust particles in the air, with the aid of dust sensors such as the digital micro measuring instrument for respirable dust, produced by the HUND Corporation in New York, or the HAM Aerosol Monitor produced by Casella London Ltd., Milton Keynes, England. The dust sensor operates in accordance with the principle of stray light measurement at an angle of 70°. The unit is calibrated with reference to gravimetrical standard principles of measurement (see also Praml *et al.*, 1986).

Chapter 6
Mental and emotional stress

In the preceding chapters we have described the physiological implications of the general stress reaction. It has been pointed out that the
endowed hypothalamic response to almost any kind of sensory input
which may be interpreted by the organism as a challenge to its
biological harmony, triggers profane physiological and biochemical
alterations in the body; for reference, see Mason (1968); Francis (1979).
Apparently, the sole purpose of all these alterations is to prepare the
organism to meet the challenge and to respond to it in a physical
manner. From a survey of the nature of the physiological autonomic
alterations brought about by the stress reaction, one is tempted to
conclude that the purpose is to put the organism in readiness for a
physical response to the challenge, i.e. a defence reaction. Evidently,
Cannon, some 60 years ago, realized that it was necessary for the
organism to be able to find a physical outlet for this biological alarm
reaction, in the form of physical activity. Failure to do so, he thought,
might lead to mental or emotional disturbances (Cannon, 1929).
Already in Cannon's time it was recognized that the state of stress
caused a speeding up of the coagulation of the blood. This was
interpreted as a desirable mechanism for stressed animals to prevent
them from bleeding to death in case they were wounded during the
ensuing fight or flight. It now appears that this enhanced stress-induced
coagulation tendency is due to increased adrenaline secretion into the
blood, brought about by the stress mechanism (Hinkle, 1973). It is
generally recognized that this increased coagulation tendency may be
particularly undesirable for individuals who suffer from ischemic heart
disease.

Studies in Squirrel monkeys subjected to environmental stimuli have
indicated that stress can induce sustained arterial hypertension (J.A.
Herd, 1972). In our own field studies of mentally stressed sea captains,
we have observed a prolonged elevation of the blood pressure (Figure
6.1). In fact, it appears that the blood pressure is even more sensitive to
stress-stimuli than is the heart rate. The stress-induced blood pressure
elevations tend to last longer than do the heart rate reactions. As we
shall see in subsequent chapters, a persistent elevation of the blood

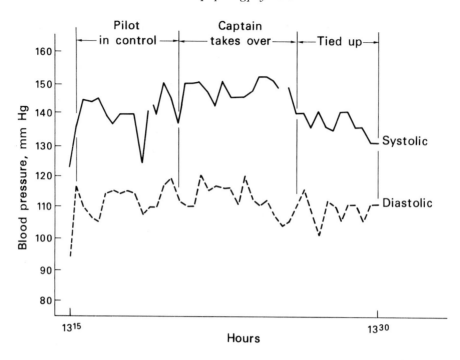

Figure 6.1. Blood pressure in the captain in command of a coastal steamer entering the port of Sortland in Northern Norway.

pressure has been observed in employees of a company threatened by bankruptcy (Rognum and Erikssen, 1980), and in air traffic controllers during a conflict concerning wages and working conditions (Mundal *et al.*, 1983). In the case of the air traffic controllers (see Chapter 9), it was not a matter of losing their jobs, but of being highly unpopular with the community and especially the airline passengers for obstructing the normal flow of airline traffic. In both the above mentioned stress situations, which lasted only some months, the levels of blood pressure in all cases returned to normal as soon as the stressful situation was over. This was not the case, however, in a third instance, involving the employees of a large ferro–alloy production plant, subject to persistent rumours of having to be closed down due to failure to make the plant profitable. Here a general elevation of the level of the individual blood pressures of the employees was observed during the first year of the threatening crisis, and was found to persist during the second and third year of the continued threat of having to close down (see Chapter 10).

It is interesting to note that Alvarez and Stanley (1930) almost 60 years ago reported significantly higher mean blood pressures in the prison guards than in the prisoners in American prisons, without associating this in any way with the stress of work.

Jonsson and Hansson (1977) have shown that prolonged exposure to a stressful stimulus in the form of noise may cause raised blood pressure in humans. They reported that systolic and diastolic blood pressures were significantly higher in 44 male industrial workers with a noise-induced auditory impairment, than in 74 males of the same age with normal hearing. Moreover, significantly more individuals with hypertension (resting recumbent blood pressure over 160/100 mm Hg) were found in the group with noise-induced loss of hearing. In accordance with this finding, A. Smith (personal communication) of the ferro-alloy company Trollhätteverken Ltd. in Sweden, observed higher incidence of hypertension in the workers with a noise-induced hearing loss. Among the 1,321 industrial workers under his care, 80 were hypertensive (6.1 per cent). In those with unimpaired hearing only 2.3 per cent had hypertension, as against 20 per cent in those who had a 50 per cent hearing loss.

It has been known since the studies of Cannon that one of the most sensitive targets for emotional or mental stress is the gastro–intestinal tract (Davenport, 1981). Studies in monkeys exposed to stress with which they were unable to cope showed that the monkeys eventually developed peptic ulcers. It is also a common experience that ulcer patients often experience aggravation of their symptoms and signs of ulcer when exposed to stress. The observations by Wolf and Wolf (1943) on their patient Tom, who had a gastric fistula, indicated that they could distinguish between the effect of anger, which increased gastric motility, acid secretion and vascularity, and fear, which had the opposite effect.

It thus seems likely that a variety of stressful situations in modern life may evoke the perception of a threat to our physiological homeostasis, which in the long run may cause organic changes which may turn out to be pathological in nature. It is also possible that this is so because of our inability to find a physical outlet for the state of organic tension created by the stress reaction, as originally suggested by Cannon (Herd, 1972).

According to Levi (1968), similar hormonal responses may take place whether the stress is in the form of an unpleasant incident, or in the form of a highly amusing experience. The urinary elimination of catecholamines in subjects watching different kinds of movies showed that the catecholamine excretion rose significantly both during the very tragic and agitating movie *Paths of Glory* by Stanley Kubrick, and the charming and highly amusing film *Charley's Aunt* by Hans Quest, as compared to watching a blend of neutral natural-scenery films produced by the Swedish National Railway Company (Levi, 1968). It thus appears that while a similar stress reaction in terms of catecholamine excretion may be elicited by a pleasant and by an unpleasant stress situation, the organic effect may be positive in the former case, but not in the latter. This may be the factual basis for the common assumption that only stresses with which we are unable to cope are harmful. It is also conceivable that at least part of the difference lies in the fact that amusing situations, although stressful, may give rise to dynamic physical activity, such as vigorous laughter or other forms of muscular engagement, which serve as a form of physical and emotional outlet for the stress-tensed organism, as originally conceived by Cannon (Herd, 1972). There is no doubt that hormones can greatly disturb the internal milieu of the organism, particularly if no muscular exertion is allowed to take place (Hinkle, 1973).

The profound physiological consequences of laughter, which is a unique human phenomenon (Darwin, 1890), may well have considerable stress-related implications, especially when it is vigorous or excessive (Black, 1984). Apart from its emotional significance, its physiological implication, when vigorous, includes engagement of the intercostal muscles, as well as the diaphragm. The latter causes the exertion of rhythmic pressure changes on the abdominal organs, especially the liver, stomach and the colon. This may, in turn, have a significant effect upon the blood supply to these organs, as well as on the peristaltic movements of the guts. And, besides, in the words of Darwin (1890): 'During excessive laughter, the whole body is often thrown backwards and shaken,...'

In the following, we shall present a series of examples of stress, based on data collected during a variety of studies of individuals from different walks of life and occupations during the performance of their work.

Examples of the stress of getting to work on time

A great deal of stress may be encountered in the process of getting to work, especially in the case of families where both parents work, and where children have to get to school on time. The simple task of getting everyone out of bed in time may be a stressful experience, not to speak of the problems involved in everyone getting into the bathroom for the accomplishment of the personal necessities in time to catch the waiting transportation.

One of the most striking examples of stresses of this kind was a commercial aircraft pilot flying one of the small twin-engine Otters on scheduled flights along the northern coast of Norway. He had chosen to live in the country, a considerable distance away from the airport, where he had to report at 0400 hrs in the morning for an early scheduled take-off. During the winter it was, at times, a major effort to get to work on time, due to heavy snowfall during the night and the possibility of being stuck in deep snow on the road to the airport. This frequently caused the pilot to lose a lot of sleep, an example of which is illustrated in Figure 6.2. In contrast, this reaction was absent in the co-pilot who experimentally spent the same night at the airport hotel, prior to take-off, just to show the difference (Figure 6.3).

Similar problems of stress associated with getting to work on time were observed in air traffic controllers living far away from their place of work, the airport, especially during the winter when the roads were icy and slick, or unplowed after a heavy snowfall. Even though the employer had provided overnight facilities at the airport for the air traffic controllers, most of them preferred to take the chance of getting to work on time, and spent the night in their own bed at home.

An even more drastic example of the same problem was observed in a radio news broadcasting employee who had chosen to live in a remote village in the hills far away from the broadcasting station, at the expense of having to drive by automobile some seventy miles over the mountains in order to appear at the broadcasting studio in time to be on the air at seven o'clock in the morning. He started the journey rather relaxed, with a heart rate of 75 beats per min. As he was approaching the city, and catching up with an increasingly heavy traffic, his heart rate rose to a level of about 100 as a sign of an increasing fear of being too late.

Sometimes the driving conditions alone on the way to work may be a source of mental stress, as is evident from Figure 6.4, which shows the

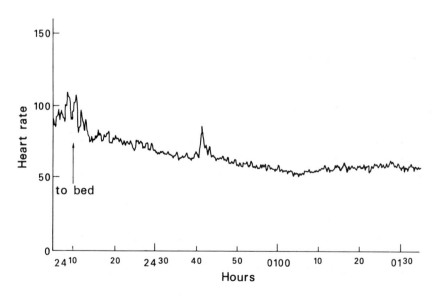

Figure 6.2. *Heart rate in a commercial pilot, worried about getting to work on time due to a snowfall in the night.*

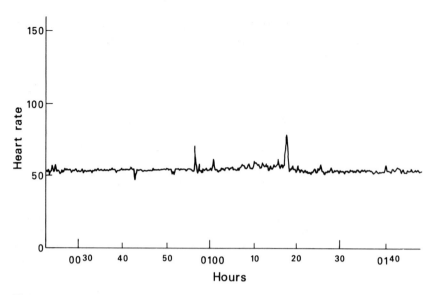

Figure 6.3. Heart rate in a commercial pilot (co-pilot to the subject in Figure 6.2), who spent the night prior to take-off in a hotel near the airport.

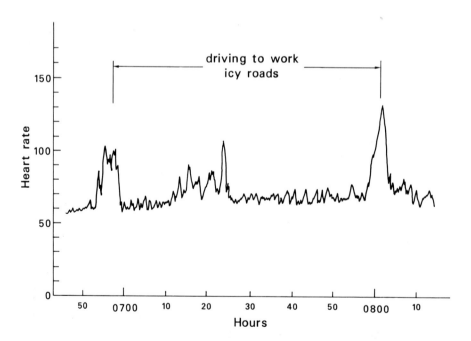

Figure 6.4. Heart rate reaction of a managing director to icy roads when driving to work in the morning.

heart rate of the general manager of a small manufacturing company encountering a stretch of icy roads on his way to work.

Examples of stress at the place of work

Needless to say, a major proportion of the stress encountered in everyday life is experienced at the place of work. The specific aspects of the stress of work will be dealt with in greater detail in the chapters that follow. Here we shall simply present a few examples of stress at work of a more general nature, which may not necessarily be associated with any one particular kind of occupation.

An example of the effect of apprehension and insecurity associated with lack of experience is presented in Figure 6.5, which shows the stress reaction of an inexperienced bank clerk involved in the process of adding up and balancing her accounts at the end of the day, compared to an experienced clerk, for whom this was strictly a matter of routine. From Figure 6.5 it is evident that the inexperienced clerk is getting increasingly stressed as time goes on, as she is having some difficulties in getting the figures to balance.

Figure 6.6 shows a bank clerk involved in the routine service of bank customers. She has developed the habit of grasping the opportunity, at rather regular intervals, of getting up and out of her seat to collect different papers or forms, or to place certain items in the safe. In so doing, she obtained a dynamic work rhythm in which her body position and work load could be intermittently altered for optimal distribution of work and rest, and for changing the load on the different parts of her body.

A commercial pilot, flying an airplane full of passengers, suspended in thin air, obviously is aware of the burden of his responsibility, and the consequences of his decisions. However skilled and experienced he may be, when faced with a critical situation, his heart rate, as a rule, reflects his state of stress. Figure 6.7 shows the heart rate of a pilot trying to land a twin-engine Otter filled with passengers, at a small air strip alongside a steep mountain with strong crosswinds causing a great deal of turbulence. His first attempt is a failure; he has to make another pass, and the second time he succeeds in putting the aircraft safely down on the ground.

Figure 6.8 shows a pilot in a similar plane, trying to land at another small airport in amongst the mountains under difficult weather conditions with thick fog and bad visibility. He makes two unsuccessful attempts before going on to the next airport. As is evident from Figure 6.9, his co-pilot, who is not flying, is apparently unaffected by the difficulties, as judged by his heart rate.

In view of the fact that practically no physical effort was involved in

(a)

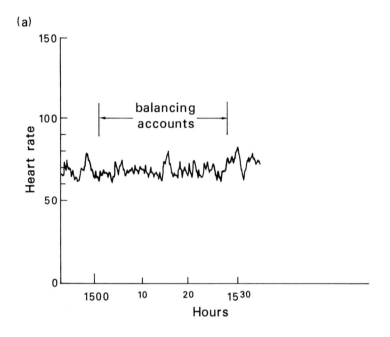

(b)

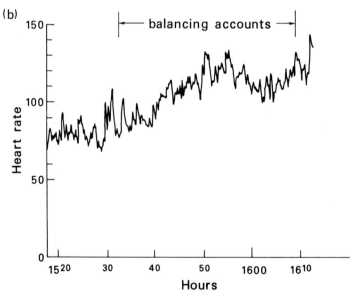

Figure 6.5. Stress reaction of an inexperienced bank clerk involved in the process of balancing her accounts at the end of the day (b), compared to an experienced clerk (a).

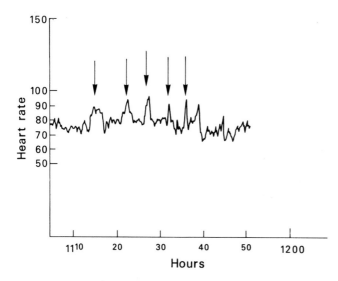

Figure 6.6. The rhythmic work routine of a bank clerk, getting up at fairly regular intervals to collect papers and forms, etc. (indicated by arrows). The result is a dynamic work rhythm providing change of body position and circulatory variations.

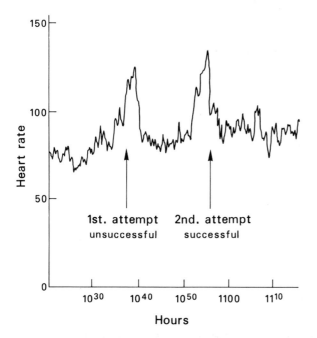

Figure 6.7. Heart rate response in commercial pilot landing a Twin-Otter aircraft at a small airfield under difficult weather conditions. He made one unsuccessful attempt but succeeded to land on the second attempt. He is not performing any actual physical work during the landing.

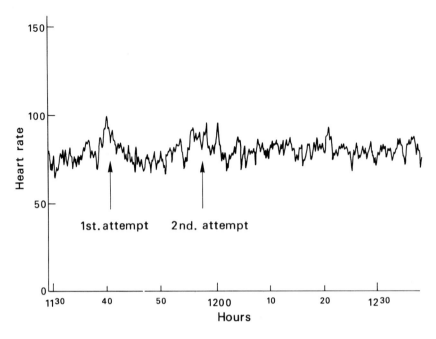

Figure 6.8. Heart rate in a pilot of a commercial Twin-Otter aircraft, making two unsuccessful attempts to land at a small airport in amongst the mountains under difficult weather conditions.

any of the above flight situations, it appears safe to assume that the elevated heart rates observed are largely due to mental stress. This is in contrast to the conclusions drawn by Roscoe (1978) who maintains that in the case of fighter pilots, flying is work not stress.

Besides emotional stress, pilots of small commercial aircraft may also be subject to considerable thermal stress, as is evident from Figure 6.10, which shows large fluctuations in the temperature in the cockpit during flight and on the ground.

Foreign exchange traders, buying and selling foreign currency on the open market for their banks, are generally considered to belong to a rather stressful occupation. A visit to such a trading centre may give the appearance of a great deal of stress and anxiety. Judged by their heart rate response, however, this may not necessarily be the case. In our experience, their stress response, as evidenced by their heart rate, is not materially greater than what is observed in many types of office workers. At one point, however, one of the trader's heart rate indicated some degree of stress, when he had bought ten million dollars and only sold six, and he was anxiously awaiting the establishment of the new exchange rate of the dollar. Similar findings were made when studying a bank employed trader at the local stock exchange centre. In this case

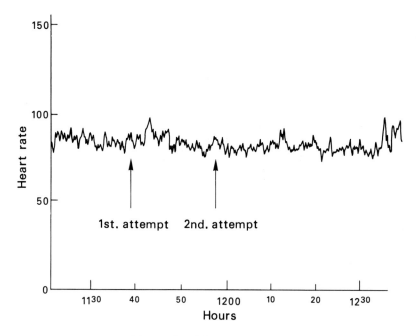

Figure 6.9. *Heart rate of the co-pilot of the same aircraft as in Figure 6.8. He is not flying and appears to be unaffected by the difficulties.*

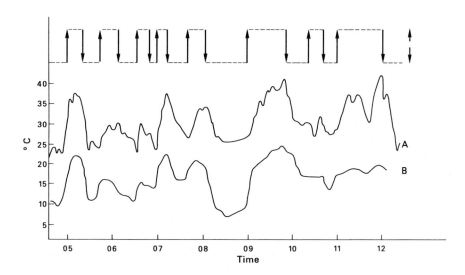

Figure 6.10. *Air temperature in the cockpit of a Twin-Otter passenger plane at floor (B) and head (A) level. Arrows indicate take off (↑) and landing (↓).*

most of the trading was done on behalf of other customers, with limited personal involvement. This may explain the rather moderate response to the outcome of the trading in terms of heart rate elevation.

An example of the emotional involvement of a general manager when addressing his employees on the subject of the need for increased productivity in the plant for the sake of greater profit is shown in Figure 6.11.

The effect of a simulated bank robbery on the heart rate of a bank cashier is illustrated in Figure 6.12. The bank employees were informed that a simulated bank robbery, including the taking of hostages, would take place at the closing time of the bank, at 3 p.m., on a certain day. One of the cashiers was chosen as a subject for the recording of the heart rate as an indication of the stress experienced. The ECG-electrodes were attached and connected to a Squirrel 1201 meter/logger in advance. At 3 p.m. precisely two masked and armed 'robbers', one with a pistol and one with a short-barrelled shot-gun, came rushing in through the main door, shouting their orders to those who were present in the bank. As is evident from Figure 6.12, our subject reacted by a moderate increase in his heart rate at this opening incident, which as mentioned, was an expected happening. All the bank employees were then ordered to lie flat on the floor, face down. This caused the heart rate to drop. Then one of the 'robbers' discovered the heart rate logger

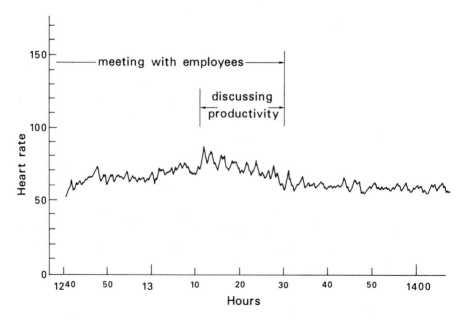

Figure 6.11. Heart rate reaction in a general manager during a discussion with his employees in a general assembly on the need for greater productivity.

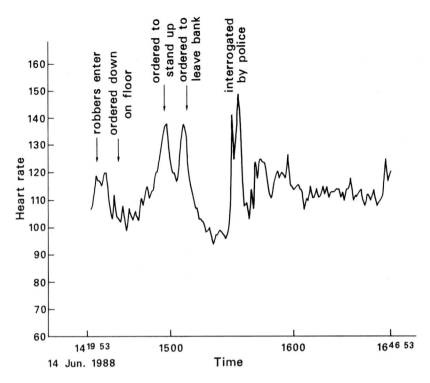

Figure 6.12. The heart rate reaction in a bank teller during a simulated bank robbery, including the taking of hostages. The time for this simulated exercise was known to all the bank employees in advance.

mounted on our subject. This caused a considerable amount of turmoil, which evidently was unexpected by our subject, who was ordered to stand up for a closer physical examination. This resulted in a sharp rise in our subject's heart rate, as was also the case when he suddenly and unexpectedly, was guided to the front door and thrown out of the building, facing the crowd of spectators, and the police force that had surrounded the building. Our subject was then interrogated by the police. This caused another sharp rise in heart rate, which remained elevated for some time.

This simple experiment clearly indicates that even when the subject is properly informed beforehand about an incident of this kind, and is aware of the fact that it is not real, and therefore harmless, it represents a surprisingly great stress, as judged by the heart rate reaction.

Examples of stress at home

Our experience from more than twenty years of stress research indicates that a major proportion of the emotional stress experienced in present-day society is rooted in problems at home. In all probability some of the stress, which individual workers become aware of at their place of work, may actually stem from problems at home carried along subconsciously in the individual's mind.

It also appears that women are more stressed at home than men, especially when they work full-time in addition to their many jobs at home. It appears that the vast variety of problems among members of the immediate family, as well as those of the more distant family members and in-laws, are apt to involve the woman more than the man. This, in addition to the many problems of her own, may be more than she can handle.

The fact that in so many families the living standard, and the cost of maintaining this standard, is so high that a dual income is necessary in order to make ends meet, is obviously one of the main reasons for the mounting stresses of modern society. A major consequence of this is less energy left over to be devoted to meaningful family activities at home. Consequently, a considerable part of the leisure hours is spent in front of the television. This may be a healthy occupation as long as the program is amusing, especially if it provokes vigorous laughter. But the watching of thrillers such as the movie *Coma* may cause sustained tension, both physical and mental, as illustrated in Figure 6.13. In this case, the heart rate remained elevated throughout the entire film, with two peaks during incidents which were particularly frightening. To what extent such regular exposure to mental tension combined with peripheral muscle tension may represent a health hazard remains to be seen.

Figure 4.6, in Chapter 4, illustrates another type of stress which may

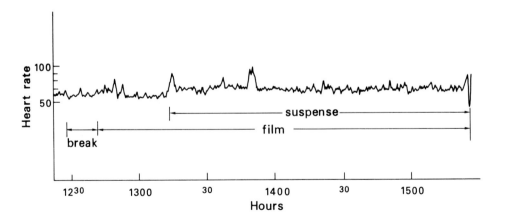

Figure 6.13. Heart rate in a young woman watching an engaging movie.

be far more common than generally recognized. This is a case of an engineer on board a modern container vessel en route from Singapore to Hong Kong. The heart rate recorder was mounted in our cabin laboratory. The engineer climbed the one flight of stairs up to his own cabin, undressed, went to bed, read for a while, became annoyed at the slamming of one of his cupboard doors, got up to close it, put out the light, and tried to go to sleep. Then he became conscious of a problem which he had brewed on subconsciously since Singapore, when he received a letter from his wife informing him that his automobile insurance company had refused to cover the expenses involved in repainting his car after the repair of damages inflicted on it during a collision the day before he sailed. This unpleasant fact bothered him now. There was nothing he could do about it, being so far away. Yet he could not forget it. The more he thought about it, the more it annoyed him. Not only the fact that he had to pay for something his insurance company, in his opinion, certainly should cover, but equally the fact that he was unable to get it out of his mind so that he could go to sleep. Eventually, several hours later, he did go to sleep.

This may be a typical example of subconscious stress appearing at the conscious surface at night, often at the lower end of the circadian scale around 4 o'clock in the morning, depriving the victim of valuable sleep.

Chapter 7
The stress of management

Most studies concerning occupational health, including work physiology, have been done on workers. The industrial leaders, obviously, also have their problems, and may be exposed to a considerable amount of stress (Brief, 1980). In the past, most of the interest in this respect has been focused on the large companies, where the leaders are recruited from a selected population, presumably well qualified to cope with the stresses to be encountered in their leading role. This may not be the case, however, in small industrial operations, especially in a small country like Norway, with limited and scattered professional and technical resources of the kind industrial leaders may need. In some cases the industrial operation may even be the result of an individual having an idea, which has been successfully put into practical application, resulting in the manufacturing of an industrial product. In the cases when the individual in question does not have the necessary commercial training and administrative experience, this may be problematic. It may be particularly difficult if the individual and his organization are located in an outlying district, where professional and technical support may not be readily available. In a country like Norway, with a sparse and highly-scattered population, combined with a lack of natural resources, individual enterprises of this kind may play an important role in forming the basis for settlement and continued habitation of outlying areas. In Norway, small- and medium-sized industrial companies employing up to 150 people, constitute about 95 per cent of all the industrial companies and employ about half of all the working force in the country.

These were some of the reasons why, between 1982 and 1985, we undertook to investigate a representative group of leaders of small industrial companies from different parts of the country (Rodahl *et al.*, 1985). Since it is not usually possible, in the case of industrial leaders, to separate work from family and social life, it was decided to focus attention on the leader's total life situation, including his state of health, and his stresses of all kinds, mental as well as physical. Previous studies in the USA indicated that managing directors of large companies were, on the whole, healthier than most people. We therefore considered it to be of interest to examine this aspect as well for the leaders of small and medium-sized companies, since it appeared conceivable that their situation might be different, compared with the leaders of larger companies.

In Norway, the term 'small and medium-sized companies' usually means a company that has less than 150 employees. Corresponding foreign companies with which they may be competing, usually employ up to 250–300 people, and they have access to greater resources and people with a wider spectrum of competence. In contrast, the Norwegian small and medium-sized companies are as a rule too small to afford a proper staff of specialists to cover the different aspects of the company's functions, including technical, financial, legal and administrative aspects. As a result, the manager alone, or a few individuals, may have to handle a variety of jobs simultaneously and without adequate help.

The available literature dealing with specific studies of the stress of work and the health of leaders of small and medium-sized companies, is very limited. Certain occupational health aspects of the leaders of small companies are dealt with in a supplementary issue of the *Scandinavian Journal of Work, Environment and Health* (Tolonen, 1979). Cooper and Melhuish (1980) examined 196 British industrial leaders, with particular emphasis on the health-related consequences of their stress of work. They were all industrial leaders who had attended the Administrative Staff College in England. This institution educates the top bracket of British industrial leaders. Thus, as the above mentioned study deals with a highly selected group, most of them employed by large companies, the group may therefore not be directly comparable to the leaders of small and medium-sized companies in Norway. Mjøs *et al.* (1982) have described the everyday life of the typical Norwegian small company leaders on the west coast of Norway, and Mjøs and Jevnaker (1978) studied co-operation among 50 small industrial companies in a community called Osterøy on the west coast of Norway. They pointed out that extensive informal co-operative arrangements existed between the leaders of the small industrial environments in the community. According to Sørlie (1982) most of the professional information reaching these small companies was the result of personal contacts among the company managers.

In our study, a total of 21 Norwegian male managing directors of small- and medium-sized companies were selected according to a table of random numbers (Rodahl *et al.*, 1985). The study included a systematic 24-hour field survey at the subject's place of work, extending over a period of 1–2 days, with a detailed review of his professional activity, his job-related problems, and how he himself experienced his work and life situation. This was supplemented by the keeping of a detailed activity log, combined with the continuous recording of the manager's heart rate over a 24-hour period as an indication of his stress reaction. This was done with the aid of a portable, battery operated ECG tape recorder. In addition, the manager was subjected to a comprehensive medical examination, the purpose of which was to gain an insight into his state of health with particular emphasis on symptoms and signs of coronary heart disease, gastro-intestinal disease, and psycho-social problems, as well as his sleep and exercise habits and risk factors such as tobacco smoking, overweight, high blood pressure and blood lipid levels. The protocol for the medical examination was identical to the one used by Erikssen *et al.* (1981) in their study of 2,114 presumably healthy Norwegian men. The study also included a standard questionnaire providing information about the manager's work and leisure, with particular reference to stress, state of health, medical history, smoking habits, alcohol consumption, exercise, etc. This was followed by a thorough personal interview to expand on the questionnaire findings, the purpose of which was to gain a further insight into how the manager himself experienced his work and life situation.

The standardized medical examination was done at the Oslo City Hospital and at the Institute of Work Physiology in Oslo. It included the examination of blood and urine, as well as an X-ray examination of heart and lungs, and a work-ECG examination.

Of the 21 managing directors examined, all but one were above 40 years of age. Eighteen of them were married, one was divorced, one was a widower, and one was unmarried but lived with a female companion. Five of the subjects were employed by the company they managed, 16 were sole or part owners of the company, nine of them having inherited the company from their parents. Seven of the 21 subjects had founded the company which they were now managing. Five of the 21 managers said that they would not recommend their children to follow in their footsteps. An equal number said they would.

Personal interviews

The personal interviews revealed that 18 of the 21 managers enjoyed their job and their role as leaders. Most of them looked forward to going to work in the morning. One of them felt that in his experience it was preferable to be a salaried director of a large company in view of the fact that this would allow him to take more time off for leisure. One of our subjects had decided to give up his leading role, and two of them would have chosen a different career if they were to do it over again. Some of those who owned their companies felt that it was not necessarily an advantage to own the company they worked for; they felt like employees anyhow, they said. In fact, some of those who had invested all their own financial resources in the company felt that they

were tied to it; they could not simply take their hat and leave. It was
pointed out that it may take a particular kind of personal ability to get a
company started, but entirely different kinds of abilities to keep it
going, once started. It was a typical finding that so many of the
managers were looking forward to being relieved of much of their daily
responsibilities and to have more time for long-range planning.

The kind of problems which were said to cause most of their stress
were, in half of our subjects, problems related to their employees,
ranging from irritation caused by inadequate performance, to major
conflicts in connection with the necessity of having to fire associates. A
few of the leaders, however, failed to see that they had any problems,
while their wives pointed out that indeed they did.

The feeling of not being able to devote sufficient time to their
families, and too much absence due to travel, were major stress factors
for many of the leaders. Most of them complained of too much official
red tape, and too much paperwork imposed on them by the authorities.
'Too many deadlines' was a common complaint. A few of them felt that
they had some difficulty in being able to steer the development of their
companies. A couple of them were constantly fighting for the survival
of their companies, having difficulties paying their running expenses.
One of them had invested a great deal of money in sophisticated
equipment which did not work properly. In the outlying districts it was
difficult at times to recruit competent key staff members with the
necessary level of professional education. The problem of keeping
capable co-workers in key positions was, at least in one case, a problem.
In several of the subjects, job-related problems were at times the cause
of sleep disturbance.

The need to expand in order to survive was, in one case, a problem.
The fact that the equipment and machinery used in production were
constantly becoming more efficient and more expensive caused a
dilemma: in order to pay for it, one had to produce more. In order to
sell the products the market had to be expanded. From a rather
comfortable solo existence on the initial local market, one had to
compete on the larger national market, and eventually on the international scene with all the added burdens this entailed. Thus, unintended,
the company had no choice but to grow.

Although the personal economy of the industrial leaders we examined was on the whole quite good, most of them took less vacation than
the rest of the employees of the company. Their alcohol consumption
was not excessive. Most of them exercised regularly, and underwent
regular medical examination, and had little or no absence from work
due to illness. Most of them were actively engaged in community and
business organizations.

Generally speaking, the subjects studied may be divided into three
categories: (1) Those who had taken over a family business, (2) those

who were employed by an already existing company, and (3) those who had started and developed their own company from scratch. In the latter category, four of the subjects started with a practical idea, and before they knew it they had a company to run without having the formal training and competence to cope with all the kinds of problems which this entailed.

Of the 21 managers examined, two of them said that they never felt stressed, although the wife of one of them made it clear that her husband at times was certainly exposed to considerable stress. Whether he himself was not aware of it, or did not want to admit that he was stressed, is an open question. Of the remaining 19 managers, 13 said that they seldom felt stressed, while six of them said they often felt that they were under stress. Only one of the 21 managers felt that he was greatly bothered by the stress he was exposed to in the performance of his job.

The most frequent stress symptoms were restlessness and inability to concentrate on their work, and irritability. Less frequent were sleep disturbances, emotional sweating, dyspepsia, headache and muscular tension.

All but two of the 21 managers examined said that they exercised regularly for the sake of physical fitness, six of them at least five times a week.

Heart rate and time-activity logging

Detailed registration of the subject's activity throughout one or more 24-hour periods by a time-activity log kept by the subject himself were compared with continuously recorded heart rate as an indication of stress response. This provided a revealing picture of what they did, what stress they were exposed to, and how they reacted to it.

This heart rate and activity logging revealed that during the actual study period, only eight of the 21 managers engaged in any physical exercise to speak of. On the whole, however, the information obtained during the personal interview agreed quite well with the facts recorded on the activity and heart rate logs. It was evident that those who did participate most regularly in physical training had considerably lower resting heart rates than those who did not.

A main impression both from the time-activity log and from the personal interview was the wide variety of jobs most of them had to perform during their daily work, and that they worked long hours.

It is evident from the heart rate records that the subjects reacted by an increase in heart rate, both to 'positive' stress such as hearty laughter, as well as to 'negative' stress, such as driving a car on icy roads (Figure 6.4, Chapter 6). They reacted with an accelerated heart rate to physical

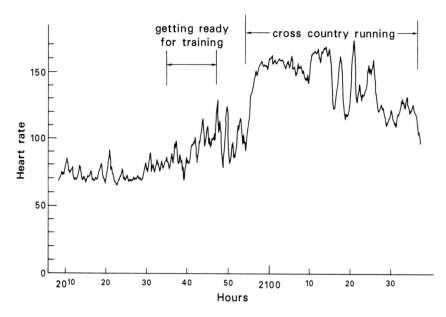

Figure 7.1. Heart rate reaction to physical training in a managing director of a small company.

as well as to mental activity. Most cases of elevated heart rates were the result of moving about physically, especially climbing stairs, etc. As to be expected, by far the highest rise in heart rate was observed during physical training, such as running, skiing or gymnastics (Figure 7.1). Another typical observation was the elevation of the heart rate in all cases when any of our subjects participated actively in meetings, by giving talks, taking part in discussions or acting as chairman. In some, this state of elevated heart rate lasted even during breaks between performance, indicating a persistent state of tension. This was the case in internal company meetings, as well as in public meetings of political, professional or other organizations. Improvised, not planned or antic- ipated, routine briefings of occasional groups of individuals visiting the company or plant, on the other hand, appeared not to cause any increase in the heart rate. Mental and emotional engagement caused, in some cases, a marked increase in heart rate, as in the case of an intense and emotional discussion of the problems facing modern industry in a small group of interested but non-opposing listeners (Figure 7.2), and a discussion of productivity during a general assembly of all the employ- ees of the company (Figure 6.11, Chapter 6). Even discussions with his associates in his own office at times caused noticeable heart rate elevation, indicating mental tension on the part of the manager. Several of our subjects reacted to unpleasant telephone conversations by increased heart rate, and by unpleasant incidents of a private nature, as

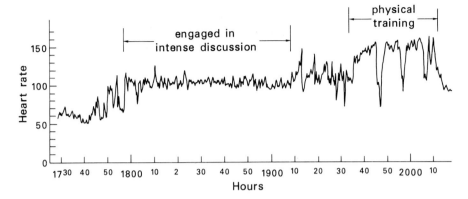

Figure 7.2. Mental and emotional engagement causing a marked increase in heart rate in a managing director engaged in an intense discussion of the problems facing modern industry in a small group of interested, but non-opposing listeners.

in one case when the manager's daughter was involved in an automobile accident. On the other hand, it caused no increase in one of our subject's heart rate when driving his car he happened to see his divorced wife in the street. Apparently, most routine office work appeared not to be very stressful. Occasionally, some of the subjects might wake up in the middle of the night and stay awake for prolonged periods thinking about problems at work, without this necessarily affecting their heart rate.

Medical examination

The results of the medical examinations revealed that four of the subjects were overweight, and one was underweight. Two were under treatment for hypertension, one had diabetes. Four of the managers smoked 5–25 cigarettes a day.

ECG at rest showed one case of definite sequelae of cardiac infarction. In two additional cases the findings were suggestive of past cardiac infarction. As expected, the signs of cardiac ischemia became more pronounced during exercise. Sixteen of the subjects had normal work-ECG. One subject developed cardiac arrythmia at his maximal work load, which disappeared as soon as the exercise was discontinued.

The testing of their maximal physical work capacity showed, on the average, superior values compared to the mean values for healthy men of corresponding age, tested by Erikssen *et al.* (1981a), confirming the information given in the questionnaire concerning habitual exercise and physical training. Resting heart rate, maximal heart rate, and systolic and diastolic blood pressure were, on the average, quite similar to those reported for healthy Norwegian men of similar age by Erikssen *et al.* (1981a).

Pulmonary function tests showed normal values. X-ray examination of heart and lungs showed normal findings, except in one case which showed possible evidence of asbestos exposure.

Routine blood tests showed normal findings, except increased sedimentation rate in one case, and one case of elevated blood cholesterol levels and moderately elevated triglyceride values in three cases. Tests of liver and kidney functions showed normal values. Routine examination of the urine was negative in all cases.

The general impression of this group of managing directors of small industrial companies in Norway examined in this study is that they, as a group, are rather healthy, fit and industrious. Although this study only included a small group of 21 individuals, it is our impression that the findings may be fairly representative of this category of industrial leaders, as a whole, in Norway. Whether or not some of these findings may be representative for managing directors of small industrial companies outside Norway remains to be seen.

Chapter 8
Shift work, circadian rhythms and transmeridian dyschronism

The fact that some of the basic physiological functions which determine our performance capacity are subject to circadian rhythmic fluctuations, reaching their lowest level during the night, supports the idea that Man is made to work during the day.

Circadian rhythms

A number of physiological parameters, such as oxygen uptake, heart rate, body temperature, and urinary excretion of potassium and catecholamines, show distinct rhythmic changes in the course of a 24-hour period, with the values falling to a low dip around 4 a.m., and rising during the day, reaching their peak in the early afternoon. Blood pressure has been found to be lowest at 3 a.m. and to begin to rise again during the early hours of the morning before waking (Millar-Craig *et al.*, 1978).

These circadian rhythms are believed to be regulated by several separately operating biological clocks (for references see Folkard and Monk, 1985). This is the case in most individuals, although there are apparently exceptions (Folk, 1974). In some individuals the high and low peaks of the circadian rhythm occur earlier than in the average individuals. They are the so-called Type-A individuals who prefer to rise very early in the morning, and to go to bed early at night. The so-called Type Bs are slow at getting started in the morning, but may feel fit to carry on until late at night. Östberg (1973) observed that the 'morning' types had their mean circadian oral temperature maximum five hours earlier than the 'evening' types.

It appears that the inherent regulatory control of the circadian rhythms may not be fixed at a fixed periodicity of 24 hours, but may vary roughly between 21 and 26 hours. In a study of members of an American overwintering party in the Antarctic, who spent their time in a snow-covered camp where the individual members lived isolated, and unaffected by each other, surrounded by complete darkness, the circadian cycle turned out to be 25 hours. That is, each day the subject got up one hour later than the day before, without knowing exactly what time it was. These observations have been substantiated in laboratory experiments by a number of investigators (Wever, 1985; Minors and Waterhouse, 1985).

It is evident that the internal biological clocks are affected by stimuli from the surrounding environment, external circumstances, activity, light and darkness, habit, etc. In the northern hemisphere it is often observed that in the summer, when it is light 24 hours a day, the sleeping pattern is apt to become floating. A majority of the inhabitants are active half-way through the night, and may find it difficult to reverse to the usual sleeping pattern in the fall.

Evidently, the inherent biological clocks are more or less keyed to the local time at the place where the individual is living. Changing to different time zones will accordingly upset the existing rhythmicity, and may affect the normal sleeping pattern. Generally, it may take several days before the biological clocks are synchronized with the local time. Similarly, when a person accustomed to working during the day and sleeping during the night, suddenly reverses the schedule, and goes on continuous night work (i.e. he works during the night, and sleeps during the day), it is observed that, at the onset, his body temperature and heart rate will fall in the night, as if he were sleeping, although he is working (Vokac and Rodahl, 1975, Figure 8.1). It takes several weeks

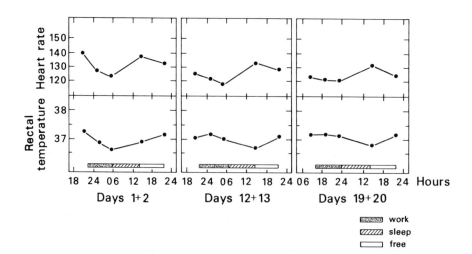

Figure 8.1. *Mean heart rate at a fixed submaximal work load, and mean rectal temperature in four steel mill workers during three weeks continuous night shifts (modified from Vokac and Rodahl, 1975).*

for this normal rhythm to be reversed (i.e. for the temperature drop to revert to an increase in body temperature in the course of the night).

The rhythmic circadian changes in physiological functions are associated with changes in performance, as pointed out in Chapter 3. This relationship appears to exist especially in the case of body temperature (rectal temperature) and performance, the lowest performance being observed very early in the morning (about 4 a.m.) when the rectal temperature is also lowest. These findings indicate that circadian rhythms should be considered when interpreting the results from prolonged physiological experiments, or when performing fitness tests in athletes at different times of the day.

Bugge *et al.* (1979) has shown that a combination of prolonged, heavy physical work and sleep deprivation causes increased fluctuation in the circadian rhythmic oscillations. Evidently, sleep is a physiological function closely tied to the circadian rhythm. Thus, Czeisler *et al.* (1980), observed in 12 subjects living on self-selected schedules in an environment free of time cues, that the duration of sleep was highly correlated with the circadian phase of the body temperature rhythm at bedtime, and not with the length of prior wakefulness.

Shift work

Ever since Man came out of the cave and ventured into a civilized kind of existence, a number of individuals have had to work at odd hours, in shifts, or around the clock, and more so in certain kinds of occupations than in others. Someone had to care for the sick around the clock. Sailors had to take watches. Bakers had to bake at night to provide the daily bread for all in the morning. Policemen had to guard our safety, soldiers had to defend us even during the night. Industrial electro-chemical processes, once started, cannot be shut down, but must be maintained around the clock. That is, those attending to the process have to work in shifts. The enormous amount of money and resources invested in industrial equipment makes it impossible to have it all standing idle for two-thirds of the 24-hour period. In fact, future trends may demand even more extended use of shift work in order for us to survive and to maintain our present standard of living.

In spite of the fact that shift work in one form or another has been practised for generations, little precise information is available as to what effects shift work has on physiological functions and on health (Folkard *et al.*, 1985). Nor is there any general agreement as to what type of shift work or work schedule is to be preferred. Much of the available information refers to clinical, social or psychological aspects of shift work (Aanonsen, 1964). A review of the literature indicates that the health of shift workers in general is good in spite of such complaints as loss of sleep, disturbance of appetite and digestion, and a high rate of ulcers. The social and domestic effects of shift work represent greater problems than do the physiological effects. The results of studies pertaining to the effects on productivity are conflicting, as are results concerning accident rates. Absenteeism because of illness appears to be lower among shift workers than among day workers.

Systematic studies of workers engaged in rotating shift work and continuous night work (Vokac and Rodahl, 1974, 1975) indicate that shift work represents a physiological strain on the organism. It causes a desynchronization between functions such as body temperature and the biological clocks governing these functions. These studies show that there are considerable individual differences in the reaction to shift work, supporting the general experience that not everyone is equally suited for such work. As judged by the catecholamine excretion, the greatest strain occurs when the worker, after several free days, starts work on night shift. The results of these studies indicate that it is preferable, from a physiological standpoint, to distribute the free days more evenly throughout the entire shift cycle. That is, to alternate between work and free time regularly instead of assigning several consecutive free days.

Different kinds of shift work

Experimental studies

In an experimental field study at a Norwegian steel mill, Vokac and Rodahl (1975) persuaded four of the steel mill workers to work continuously on night shifts for three weeks running, including Saturdays and Sundays. The purpose was primarily to find out how long it would take for the circadian pattern of the heart rate at a fixed sub-maximal work load, and the rectal temperature, to convert to the normal day-work pattern. At the onset, the body temperature and the heart rate fell in the course of the night, as if the subjects were sleeping, although they were working (Figure 8.1). At the end of three weeks continuous night shifts this rhythm was still not completely reverted. This finding was essentially confirmed by Knauth and Rutenfranz (1976), and by Knauth *et al.* (1978), in their experimental shift work studies.

These observations indicate that it takes too long for the organism to adjust to night work to make it practical to adhere to a work schedule of a long sequence of successive night shifts. Furthermore, the benefit of the reverted physiological reaction is lost when interrupted by a single day.

Fixed versus rotating shift work schedules

A great variety of different views have been voiced in the literature on shift work, on schedules for rotating shifts which may be the least harmful for the shift workers (Murphy, 1979). There are strong arguments in favour of fairly frequent changes from one shift type to another, particularly in the case of night shifts.

In the opinion of Rutenfranz *et al.* (1977), single night shifts are better than consecutive night shifts, based on the fact that a single night shift does not significantly disturb the circadian rhythm. On the other hand, it is our experience that the first night shift is usually harder to take than the subsequent ones in a consecutive series of night shifts. This is partly supported by the findings of Hawkins and Armstrong-Esther (1978), who found that short-term memory performance was significantly impaired on the first night, but improved progressively on successive nights, in nurses on night duty. It may therefore be better to take a few night shifts in a row, rather than only one at a time. It should be kept in mind, however, that there are great individual differences in terms of adjustment to night work, or to odd work hours. It should also be borne in mind that the individual, to a considerable extent, may be affected by what he was doing during the day prior to the night shift, and especially to what extent he was able to acquire an adequate amount of sleep. In our study of the steel mill workers, the first night shift was preceded by five successive free days, which by some of the workers were used for excessive activities such as house building, fishing or hiking trips, etc., associated with very little sleep, causing the worker in question to appear for work in a state of mild sleep deprivation.

Whatever the scientific facts or the opinions of the scientists may be, the actual arrangement of shift work schedules in the industry is, in our experience, most likely to be made according to what the workers themselves, for whatever reasons, may prefer or wish to vote for.

Effects on health

The health aspects of shift work have been the subject of a number of studies and debates (for references, see Rutenfranz *et al.*, 1977). Yet, the conclusions are still conflicting. Among the possible negative effects of shift work, disruption of sleep patterns and eating habits, gastro-intestinal disturbances, and nervous disorders have been emphasized. Among the most frequent complaints in shift workers are gastric and intestinal dysfunctions, while in some populations it appears that the incidence of cardiovascular disease is no higher in shift workers than in the rest of the population (Rutenfranz *et al.*, 1977). The findings of Aanonsen (1959) that those who elected to discontinue their shift work had the highest incidence of disease might indicate that the remaining shift workers are as healthy as they are because those who were unhealthy had left. In a more recent study, Knutsson *et al.* (1988) claim that shift work is associated with increased risk of coronary artery disease. A higher proportion of their shift workers smoked (53 per cent versus 39 per cent), and shift workers had higher levels of serum triglycerides than did day workers.

Taylor (1973) has critically reviewed the available literature concerning the effects of shift work on the worker's health. Underlining the magnitude of the problem, he points out that the increase in numbers of shift workers worldwide has been particularly rapid since about 1950. In the manufacturing industry in the United Kingdom, the number of shift workers rose by 50 per cent in the 10–year period between 1954 and 1964. In 1968, 14 per cent of the full-time men and 5 per cent of the full-time women were shift workers in the United Kingdom. In the United States, more than one-quarter of the workers do shift work (Baker, 1980).

Taylor (1973) points out that while shift workers in response to formal interviews claimed that they suffered from sleep difficulties, carefully conducted studies revealed that rotating shift workers actually slept for more hours than did day workers. This, however, does not necessarily mean that the quality of sleep was equally good.

A number of studies (for references, see Taylor, 1973) have shown that three-cycle, seven day rotating shift workers in continuous-process industries have less sickness absenteeism than day workers in the same firms, although other studies have come to the opposite conclusion. Taylor (1973) also points out that large-scale medical studies of shift workers failed to support the theory that shift work causes peptic ulcers. Furthermore, carefully controlled, large-scale studies have shown no significant excess mortality in men who had done ten years or more of shift work.

On the basis of his review of the world literature and studies of his own, Taylor (1973) concludes that there is little objective evidence that shift work, continued for any length of time, significantly affects health, either in terms of sickness absenteeism or in mortality. In his opinion most of the problems prove to be social rather than strictly medical – the workers being unable to adapt to the social and physiological implications of shift work.

In a retrospective cohort study comparing the state of health in terms of complaints and diseases in 270 day and 370 shift workers, carried out in a chemical industrial company, Angersbach *et al.* (1980) found that the incidence of sickness in general did not differ between permanent shift workers and day workers. However, shift workers were, on the average, sick for longer periods than the day workers. It appeared that more shift workers than day workers consulted the company health service about gastrointestinal complaints, and shift workers had more frequent gastrointestinal diseases than day workers, and more peptic ulcers. In this connection, it is of interest to note that Kolmodin-Hedman and Swensson (1975) observed a higher frequency of reported peptic ulcers in railroad engineers than in other groups. The question remains, however, whether these complaints were the result of shift work, or merely being aggravated because of the shift work. The latter possibility is supported by the findings of Angersbach *et al.* (1980) who found that gastrointestinal diseases were more frequent in shift workers with a past history of gastrointestinal diseases. Psychosomatic disorders were seldom diagnosed and did not differ between shift and day workers. Nor was there any difference between shift workers and day workers in the occurrence of cardiovascular disease, including myocardial infarction.

Another point of interest brought out by the study of Angersbach *et al.* (1980) was the fact that the shift workers were engaged in more activities during off hours, including 'moonlighting'. This serves to show that when discussing the relationship between work and health, factors affecting the individual outside the place of work must also be considered.

Although it is generally believed that gastrointestinal complaints, especially peptic ulcers, are associated with shift work, this, as already pointed out, is not supported by all the available literature on the subject. This may, at least in part, possibly be explained by the fact that many of those who are adversely affected by work at odd hours or by shift work, find something else to do, so that a majority of those who remain have little or no ill effects of shift work. This may, to some extent, explain why some of the studies have concluded that the health of shift workers is surprisingly good, and in fact indicate that there is less sickness and absenteeism from work due to illness among shift workers. This possibility appears to be supported by the findings of Angersbach *et al.* (1980), who observed that the highest incidence of sickness in general was found in shift workers who transferred later to day work, mostly for medical reasons (see also Aanonsen, 1959). On the other hand, data reported by Tarquini *et al.* (1986), on the serum gastrin and pepsinogen changes in shift workers, showed that shift work causes a marked change in the gastrin/acidopepsin secretion system, potentially harmful in terms of developing gastrointestinal disorders, including peptic ulcers.

Tasto (1982) conducted a questionnaire survey and an examination of health and safety records in about 1,000 shift workers of both sexes in the food processing industry. The results indicated that shift workers had poorer sleep, greater alcohol consumption, and greater incidence of sick absenteeism compared to the day workers. However, from all indications available, shift work does not have a significant effect on longevity and/or cause of death.

Accidents

Shift workers, particularly those on rotating and night shifts, are reported to have a greater than average incidence of work-related injuries (Tabor, 1982). It has been suggested that this may be due to lack of alertness and vigilance, and falling asleep on the job during the early morning hours. According to Krieger (1987), 60–70 per cent of the questioned American shift workers admitted that they regularly nodded off or fell asleep during the night shift. Furthermore, studies of safety near-miss events or operator error revealed that error events were most frequent at change of shift, and that night shift hours, from midnight to 6 a.m., were most severely affected (Krieger, 1987).

Andauer and Metz (1967), who studied accident rates in 11,000 workers in the mining and steel industries, found that accident rates were higher during the morning and afternoon shifts than during the night shift. On the other hand, the accidents were more serious on the night shifts.

Tasto (1982) reported a greater incidence of work-related injuries in shift workers in the food processing industry, particularly in those on rotating and night shifts, based on a questionnaire survey and on examination of health and safety records. The impact of shift work on work-related injury differed somewhat, depending on the worker's sex, age and work tenure.

Effects on sleep

Sleep is a matter of major concern in the life of shift workers and considerable emphasis has been placed on disturbance of sleep and sleep patterns as a consequence of shift work.

Maasen *et al.* (1980) registered the amount of sleep and sleep difficulties in 27 young shift workers (mean age 27–28 years) during a complete shift cycle, consisting of six morning shifts, six afternoon shifts, six night shifts and six free days (Sundays free). The mean duration of sleep during the four-week period was eight hours, the amount being significantly larger during the free week, and significantly smaller during the night shift week. They observed, as could be expected, that difficulties in getting to sleep and sleep interruptions are more common during day sleep in the night-shift period. The authors concluded that the duration and quality of sleep in these young shift workers living in a rural area was quite satisfactory. They attributed this to the shift system used, to the free week and to the young age of the subjects.

Knauth *et al.* (1980a), in a study of 1,230 shift workers, reported that the shortest night sleep was found before the morning shift. They concluded that the morning shift should not begin too early in order to avoid an accumulation of sleep deficit. However, in another study, Knauth *et al.* (1980b), examined the effect of laboratory-based experimental shift work in 20 volunteers over a total of 112 days. In these laboratory experiments with fixed sleep duration, no separate effect on sleep quality could be established for different shift systems, as judged by electroencephalograms, electrooculograms and electromyograms recorded during the sleep. It should be noted, however, that laboratory-based sleep experiments are not directly comparable to observations made in real life.

As is well known, urinary catecholamine excretion drops during sleep at night. Sudo (1980) has examined the urinary excretion of adrenaline in 15 male guards, engaged in 24-hour shift work, including a 4-hour period of sleep during the night. During the short night sleep he found that the urinary output of adrenaline was inversely related to the depth of sleep, based on a personal assessment by the subject. He found that the urinary adrenaline levels during the short night sleep were as low as the level during usual night sleep, and suggests that adrenaline output may be a useful index of sleep.

Naitoh *et al.* (1982) have presented data from 38 US Marine Corps volunteers, showing that the restorative effect of a short period of three hours' sleep from 0400 to 0700 hours in the morning, was sufficient to support subsequent mental performance for an additional 20 hours. This is more or less in accordance with the results of studies of Military Academy cadets' performance during a 5-day intensive battle course with almost no sleep. Three hours of sleep during each night caused a marked improvement of performance (Opstad, 1987). The general conclusion to be drawn from these observations is that a few hours sleep may be sufficient to restore performance even though the subject may not have had enough sleep to restore the feeling of being completely rested.

Tilley *et al.* (1982), monitored sleep and performance in 12 male shift workers, operating a discontinuous, weekly alternating, three-shift system, over the course of one complete shift cycle. Compared with night sleep, day sleep was shorter in duration and was inferior in quality. Simple reaction time and four-choice reaction time were impaired at night, and simple reaction time deteriorated as a function of the number of days into the shift. This was interpreted as an accumulative sleep deficit. For this reason, in the opinion of the authors, the best shift system is probably one having a short rotation cycle with rest days preceding and following the night shift. This, however, in our own experience, depends on what the shift worker is doing on his rest days, i.e. whether he is resting or being exceedingly active.

Anderson and Bremer (1987) examined the relationship between sleeping patterns at home and the individual's own ratings of sleepiness on the job in 29 rotating twelve-hour shift workers. They found no significant difference between early-onset and late-onset sleepers in reported sleepiness on the job. The reported sleepiness on the job was significantly less in short sleepers than in long sleepers on both day and night shifts, a surprising finding which was not entirely explained in this paper.

Åkerstedt and Torsvoll (1981) studied sleep length and subjective rating of sleep quality for 390 workers on three-shift, two-shift and day-work systems. The three-shift workers' sleep quality was best, and sleep length longest when the subjects were on afternoon shift, followed by the morning shift, and then the night shift.

Effect on performance

We all know that our level of performance varies in the course of the 24-hour cycle. As already mentioned, a study of switchboard operators long ago showed that the level of performance reaches its lowest point between 2 and 4 a.m. (Browne, 1949). A similar observation was made in locomotive drivers by Hildebrandt et al. (1974).

On the basis of a series of experimental studies on different shift work systems, Colquhoun et al. (1968a, b; 1969), concluded that body temperature was, in fact, a predictor of performance as judged by the efficiency of solving mental tasks (an auditory discrimination task and a simple addition task). Colquhoun (1970) actually studied performance at certain tasks continuously during different shifts in the laboratory over periods of 12 consecutive days, and was able to show that mental efficiency in general rather closely followed the course of the body temperature rhythm.

This is a confirmation of the findings of Kleitman (1963) who, by laboratory experiments, was able to demonstrate that reaction time was closely related to body temperature, as it changed according to the body's circadian rhythmicity, suggesting that temperature dependent changes in metabolic rate affected the speed of chemical processes in the brain.

Selection of individuals most suited for shift work

According to LaDou (1982) a careful selection of shift workers may reduce the number of health problems resulting from shift work. Folkard et al. (1979) have suggested that it may prove feasible to develop a questionnaire that would predict the degree to which an individual's rhythms would adjust to shift work, including flexibility of sleeping habits and ability to overcome drowsiness.

Reinberg et al. (1978) have suggested that a low amplitude of certain circadian rhythms might be considered as a measure of individual ability to adjust easily to shift work.

According to Rutenfranz *et al.* (1977), some 20 per cent of the working population is unable, for various reasons, to sustain shift work. For this reason, individual selection of those most suited for this type of work is indicated, based on a history of gastrointestinal and digestive tract disorders, associated with stress and odd meal times, sleeping problems and inadequate sleeping facilities.

At any rate, no one knows better than the individual himself how he feels when working different shifts. Asking him directly may therefore be as meaningful as measuring a variety of theoretical parameters.

Effect of age

With increasing age and experience of shift work, Åkerstedt and Torsvall (1981) found that sleep quality and sleep length were reduced. This, however, may not be a phenomenon unique for shift workers. In our studies of shift workers in Norway we observed that the older individuals tended to take naps in the afternoon prior to going on night shifts, thus eliminating, to some extent, the problem of sleepiness during the night.

Pros and cons of shift work

From all indications it appears that the most negative aspects of shift work are its social and psychosocial implications. The positive aspects of shift work include higher wages, more free time during the day, and several free days in succession between shifts. The main problem facing the shift worker seems to be sleep disturbances, and inability to sleep in peace during daytime for those who live in overcrowded and noisy housing facilities.

There is at least one practical conclusion to be drawn from this review of the problems of shift work: a system is necessary for the selection of shift workers in order to exclude those who are sensitive to the negative aspects of it. It appears that about one-fifth of the working population fail to adjust well to shift work. This appears to be a general phenomonen, for the same percentage of maladjustment seems to apply to many other life situations. The distribution of human characteristics, including ability to adjust to shift work, follows more or less the normal distribution curve, a fact which obviously has to be considered when trying to place a potential worker. In the case of shift work, however, one of the disadvantages is that almost all studies of shift work involve men, very few involve women.

Rutenfranz *et al.* (1977) have pointed out that religious and cultural habits among immigrant shift workers may cause some problems. As an example they mention the Moslems, who during their annual month-long Ramadan festival must fast between sunrise and sunset. Consequently Moslems on night shift during the Ramadan may have some nutritional problems.

Transmeridian dyschronism

Disturbances in circadian rhythms may give rise to considerable problems for those who have to travel by air from one continent to another in order to conduct business, to take part in political negotiations, or to participate in athletic competitions (Carruthers *et al.*, (1976). For an historical review, see Mohler *et al.* (1968), and O'Donnell (1971). It is still an open question whether the indisposition or functional disturbances experienced after such intercontinental flights are in fact due to disturbed circadian rhythms, to loss of sleep, or to both (Åkerstedt, 1985). It is a common experience, however, that by being able to sleep during such travel, the individual can maintain a reasonable functional capacity in spite of the rapid shift from one time zone to another. The problems associated with the use of sleeping pills are discussed by Preston and Bateman (1970).

Haulty and Adams made an elaborate study of phase shifts of the human circadian system and performance impairment as a consequence of intercontinental flights: east-west (1966a), west-east (1966b), and north-south (1966c). They concluded that the east-west and west-east flights caused a shift in circadian periodicity, manifested by physiological functions, evidenced by such parameters as rectal temperature, heart rate, respiratory rate and blood pressure. This was not the case when flying north-south. The north-south flight produced a significant increment of subjective fatigue, however, as did the east-west and the west-east flights. The significant impairment of psychological performance observed as a consequence of transmeridian flights was not observed on the north-south flight.

A significant performance decrement, compared to the pre-flight control, was also observed by Klein *et al.* (1970) after eastward flights. This was not observed after westward flights, however. In the opinion of the authors, this was due to an unfavourable flight schedule and more severe sleep loss connected with eastward travelling.

Winget *et al.* (1984) have reviewed the problems associated with disturbed biological rhythms (dyschronism or desynchronosis) in the aerospace environment. For further references, see Rutenfranz and Colquhoun (1978, 1979).

The problems associated with transmeridian dyschronism also affect sailors at sea moving east or west, having to set their clock back when moving westward, or forward when moving eastward, as they move from one time zone to the next. It may be a matter of changing the time one hour each day or every other day. In most cases this time change is done at night. This means that those on board who are free to sleep during the night will have an extra hour for sleep when travelling westward. When travelling eastward they will have an hour less for sleep during the nights when the time is changed. Consequently, it might be advisable to change the ship's time during the night when travelling westward, but change the time during the day when travelling eastward to provide as much time for sleep as possible for those who do not have to stand watches. For those who take watches it does not make much difference in any case.

Studies of circadian rhythm in sailors at sea have shown a lag in the biological clocks in relation to the changing local time, when moving from one time zone to the next. In a crew of a container vessel (see Chapter 13), 25 per cent of the crew of 30 complained of sleep difficulties because of a lack of synchronization between their biological clocks and the ship's time. A study on board an oil tanker travelling from South Africa to South America showed a circadian phase shift in the First Officer who worked on regular four hour watches, but not in the chief engineer, who only worked during the day and slept uninterrupted during the night (Figure 13.25, Chapter 13).

On the basis of our observations it appears that the changing of the time travelling from east to west was, on the whole, tolerated fairly well, but moving the other way causes more problems. The complaints include fatigue, indisposition, difficulties falling asleep, interrupted sleep, insufficient sleep and indigestion.

Chapter 9
The stress of air traffic controllers

Review of previous studies

The work of aircraft controllers, and its effect on their health, and on air traffic safety, has been the subject of considerable discussion. In May 1979 the International Labour Organization (ILO) arranged a conference in Geneva to discuss the problems of the air traffic controllers. Among the participants were representatives and professional advisors of various organizations concerned with the air traffic controllers and their profession. Judging from the published conclusions reached by the conference, the statements made by the participants were mainly definitions of problem areas and expressions of points of view, and recommendations pertaining to working conditions, salary, age of retirement, etc. Very little of this appears to be based on technical, medical or other scientific studies.

Crump (1979) concluded, on the basis of his review of the studies which had been made of stress among air traffic controllers up to 1979, that the stress level of air traffic controllers did not seem to be significantly different from that of the rest of the population. Nor did the shift work system, practised by the controllers, appear to be a particular stress factor, even though the available data, to some extent were conflicting. Similarly, the results of some of the studies which had been made concerning hypertension and cardiovascular disease, may be subject to different interpretations. In one study, air traffic controllers who had to handle a great deal of air traffic were found to have a greater risk of developing hypertension than a control group consisting of second class airmen. These air traffic controllers also had a higher incidence of peptic ulcers (Cobb and Rose, 1973). Maxwell *et al.* (1983), on the other hand, were unable to find any significant difference in the risk of developing cardiovascular disease between air traffic controllers and a control group. Other studies have even shown lower blood pressure in air traffic controllers compared to control groups. In one study, on the other hand, the occurrence of abnormal electrocardiographic ECG findings in the air traffic controllers was twice that of the controls (Dougherty, 1967).

The air traffic controllers themselves, in a questionnaire study, stated that they subjectively felt that they had substantially more of the kind of symptoms which are usually associated with stress than ordinary people. These symptoms included headache, gastrointestinal symptoms, chest pain and peptic ulcers. This study involved air traffic controllers who had been in service more than three years (Dougherty *et al.*, 1965).

Grandjean *et al.* (1971) measured fatigue in 68 air traffic controllers using a combination of critical fusion frequency, tapping test, grid tapping test and self-rating. They concluded that their findings were indicative of a common state of fatigue. They also observed a significant increase in the urinary excretion of catecholamines after work involving radar air traffic control in the six subjects examined.

Dr Robert Rose, then chief of the Psychiatric Department, University of Texas, directed a large-scale investigation of 416 air traffic controllers in the USA during 1969–70. In an interview with the New York Times on 9 August 1981, he claimed that about 30 per cent of all the air traffic controllers examined had elevated blood pressures and could be classified as hypertensive. In the same interview Dr Carlton Melton, chief of the Department of Physiology at the Federal Agency's Aeromedical Institute in the USA, maintained that the elevated blood pressure amongst the air traffic controllers was not primarily the result of work stress, but was mainly due to the fact that this profession appeared to attract individuals who were prone to develop hypertension. He was of the opinion that the air traffic controllers would have developed hypertension anyhow, regardless of what occupation they chose.

According to Booze (1979), hypertension and psychoneurotic symptoms are the most frequent complaints among American air traffic controllers. He did find, however, that the prevalence of hypertension and cardiovascular disease among American air traffic controllers is less than in the population as a whole. Crump (1979) also emphasized psychosocial factors and their possible long-term effect on the health of air traffic controllers. He points out that, according to the air traffic controllers themselves, what they disliked the most, was their superior's lack of understanding for, or reaction to, the problems and complaints that they raised. He refers to a study conducted at the Frankfurt Air Base which showed that the main cause of the dissatisfaction among the air traffic controllers was the administration, the pay and the working conditions, while stress was the aspect of their job which they least of all reacted to in a negative way (Singer and Rutenfranz, 1971).

Studies of Canadian air traffic controllers appear to point in the same direction, according to an article in the Calgary Herald of 3 November 1981, which referred to a research worker by the name of Arlene

MacBride. According to this source, the air traffic controllers represent a highly select group of people who can cope with their work, and who therefore are no more stressed than most people. According to Mac-Bride they consider their relationship with the management and political interference to be more stressing than any aspect which has to do with their work, such as peak periods of heavy traffic or fear of causing accidents. She even goes so far as to indicate that all the fuss about the terribly stressing work of the air traffic controllers actually may cause the air traffic controllers themselves eventually to begin to see problems in their work, which otherwise would not have existed.

Melton *et al.* (1978) compared the work stress of air traffic controllers at airports which had a small amount of traffic, with airports with very heavy traffic. They concluded that the physiological stress reactions, as judged by heart rate and stress hormone elimination, were clearly less pronounced at airports with light traffic than was the case at airports with heavy traffic. They came to the conclusion that it is misleading to classify all air traffic control work as exceptionally stressing.

In this connection it should be noted that Melton *et al.* (1978) observed no statistically significant difference in the heart rate before work and during work in air traffic controllers, except in one case where the heart rate was 11 beats per min less during work than prior to work.

Hale *et al.* (1971) examined the physiological reactions in air traffic controllers at O'Hare Airport, Chicago, which at that time was the busiest airport in the world, with up to three landings or take-offs every minute. They found that the air traffic controller's work loads were highly stressing. This caused increased urinary catecholamine elimination (adrenaline and nor-adrenaline). The peak values occurred during the busiest hours in the morning.

Melton *et al.* (1979) found that the heart rate of the air traffic controllers was significantly higher during the busy afternoon shift than during the morning shift. The control of arriving aircraft caused higher heart rate than did the control of aircraft taking off. They also observed higher heart rate levels in air traffic controllers working in the tower at the O'Hare Airport than in the tower at the less busy Houston Intercontinental Airport. A psychological study of air traffic controllers at airports with different traffic density showed, on the other hand, no significant difference in psychological stress reactions (Melton *et al.*, 1978).

Grout (1980), referring to the studies of Dougherty *et al.* (1965), maintains that the occurrence of stress-related symptoms such as headache, digestive disorders, chest pain and peptic ulcers, is not only significantly more frequent in air traffic controllers than in the population as a whole, but that these symptoms also increase significantly with the number of years in service. Furthermore, the air traffic controllers exhibit pathological ECG more frequently than the rest of the population. Referring to Cobb and Rose (1973) he concludes that air traffic controllers are more likely to develop hypertension and peptic ulcers than most people. British research workers also consider the air traffic control profession to be quite stressful (Crump, 1979).

Fowler (1980) has discussed the function of air traffic controllers in relation to air traffic safety in America, from a pilot's point of view. On the basis of the studies made by the American Safety Institute and others in connection with 'near misses', he maintains that most of the mistakes were not made during periods of intense air traffic concentration, but when the work load was relatively light, and that most of the air traffic controllers perform at their best when they are under a certain amount of stress.

Karson and O'Dell (1974) made a psychological personality survey of a large number of American air traffic controllers. They concluded that the air traffic controllers had above average intelligence, had a sense for detail, were able to be determined when necessary, were sure of themselves, and were largely free of anxiety and worries. They had, in the opinion of the investigators, the kind of abilities which were necessary to make the air traffic controllers ideally suited and proficient. Furthermore, those who were applying for admission into the air traffic control profession were even better qualified in some ways than those who already were fully qualified.

Mohler (1983), who has had extensive and prolonged association with the problems of air traffic controllers, points out that during periods of reduced visibility, air traffic controllers are the most critical factor in aircraft collision avoidance. Controllers also largely determine efficiency in the mass movement of aircraft, on instrument flight plans. Individual and group controllers' health and well-being are essential to the sustained efficient and safe operation of these aircraft in the international aerospace system. Impairment of mental function due to illness, fatigue, drugs, excessive stress, alcohol or other factors are major threats to air safety. On the basis of a review of the results of the large-scale Air Traffic Controller Health Change Study, he concludes that the population of controllers was found to have an increased rate of hypertension compared to males of similar age. He also indicates a slightly increased risk of developing peptic ulcers in the group of air traffic controllers studied.

The main points from the available literature on the problems of air traffic controllers may be summarized as follows. It is generally agreed that the work of the air traffic controller is demanding. It requires a high degree of skill, alertness and ability to handle several problems simultaneously. Yet a number of studies have indicated that the level of stress experienced by air traffic controllers does not appear to be excessive compared to the population as a whole (Crump, 1979). Nor does the system of shift work practised appear to represent a major source of stress. Health surveys among air traffic controllers have yielded conflicting results. Some studies (Cobb and Rose, 1973) have revealed a greater risk for air traffic controllers to develop hypertension and peptic ulcers than the population as a whole. Maxwell *et al.* (1983), on the other hand, found no statistically significant difference concerning the risk factors for cardiovascular disease between air traffic controllers and a control group.

On the basis of the available literature it thus appears that the findings pertaining to stress-related health problems of the air traffic controllers are conflicting, particularly as far as hypertension and peptic ulcers are concerned. Apparently, some of the air traffic controllers studied do have a higher incidence of hypertension and/or peptic ulcers than the rest of the population, while others do not. It is possible that this discrepancy, at least in part, may depend upon the circumstances at the time of the study, especially whether or not the air traffic controllers examined were involved in any major conflict causing emotional disturbances at the time of the study. This was the case in a study carried out by members of the Norwegian Institute of Work Physiology, Oslo, who surveyed the state of health with particular reference to stress in the air traffic controllers at Oslo airport, Fornebu, in 1981 and 1982 (Mundal *et al.*, 1983). The study was initiated at the joint request of the Civic Aviation Agency and the Air Traffic Controllers Union in June 1981, and was completed in the fall of 1982. During 1981, when the main study was conducted, the air traffic controllers were engaged in a rather severe conflict concerning wages and working conditions. The conflict was entirely solved and the working conditions had returned to normal before the final part of the study, which included repeated measurements of the blood pressure, in the fall of 1982.

Altogether 34 air traffic controllers, out of a total of 46 were examined, 16 of them on two, and one of them on three occasions. Half of them worked in the control tower, the other half in the control centre. The mean age of the 34 subjects was 40.5 years (29–60).

The study included a comprehensive medical examination with particular reference to blood pressure, measured by a standardized procedure, gastrointestinal disease, and risk factors such as blood lipids, etc. The subjects were asked to describe how they experienced their own work situation, using a questionnaire, followed by a personal interview. In addition, the heart rate, as an indication of autonomous nervous system response, was recorded for periods up to 24-hours with the aid of miniature magnetic tape recorders, with simultaneous recording of a detailed activity log. This enabled us to record how each subject reacted to different situations at work and during leisure hours.

The measurement of the blood pressure was made under the following standardized conditions: after a 5-min rest in a supine position, the measurement was done twice while the subject remained in the supine position. The first measurement was made for the purpose of roughly establishing the approximate level of the pressure. For the second measurement, the arm was elevated for the purpose of venous draining. In this elevated position, the blood pressure cuff was inflated up to 50 mm Hg above the blood pressure value observed during the initial blood pressure measurement. The arm was then brought down to the horizontal position, and the cuff was slowly deflated. At about 20 mm above expected systolic blood pressure the subject was asked to take a deep breath, followed by a slow expiration, and finally to hold the breath while the blood pressure cuff was deflated past the systolic point. Following this, the subject was asked once more to take a deep breath, and the same procedure was repeated for the measurement of the diastolic blood pressure.

Assessment of the work stress

The mean heart rate during work, during breaks in the course of the work shift, during travel to and from the place of work, during leisure hours, and during sleep are shown in Table 9.1. As is evident from this table, the mean heart rate was about the same during morning and during afternoon shifts, and roughly the same during work and during breaks. On the average they spent about 315 min in a working position, 90–116 min were used for breaks. They spent, on the average, 56 min travelling to and from work, and 414 min sleeping per 24-hour period. One of the air traffic controllers bicycled to work, one used the train, the rest of them went by car. Of the 20 aircraft controllers who commented on their sleeping conditions during the period of the study, eight spent the night in the special sleeping facilities provided for the air traffic controllers at the airport, two of whom said that they slept poorly. The rest of them slept at home, nine of whom complained of

Table 9.1. *Mean heart rate*

	Morning Shift	Afternoon Shift
Working	82 (61–105)	80 (64–100)
Break	84 (60–102)	84 (67–106)
Travel to and from work		84 (59–120)
Leisure		79 (61–92)
Sleep		62 (51–76)

disturbed sleep. This might indicate that the air traffic controllers, in general, slept somewhat better at the airport than at home.

A detailed analysis of the individual heart rate recordings revealed, as expected, considerable individual variations in reaction patterns. Yet, on the whole, the majority of them reacted rather moderately to the stress situations which occurred during the periods of recording. This included a near-miss situation which could have resulted in a mid-air collision between a single engine private aircraft and a commercial airliner being guided by the air traffic controller. In accordance with this, only four of the air traffic controllers stated, when asked, that they were often stressed; 23 were seldom subjected to stress, and one claimed that he never felt stressed. However, a closer examination revealed several signs of stress in one of those who maintained that they seldom felt that they were under stress. This discrepancy is to some extent supported by the impression gained from the examination of the heart rate records.

From a physiological point of view it is an open question as to whom the air traffic controllers can be compared with in terms of demand for professional skills and stress exposure. The air traffic controllers themselves have, at times, compared their work with that of the airline pilots. For this reason it may be of interest to compare the reaction of the two pilots in a small twin engine Otter aircraft attempting to land at

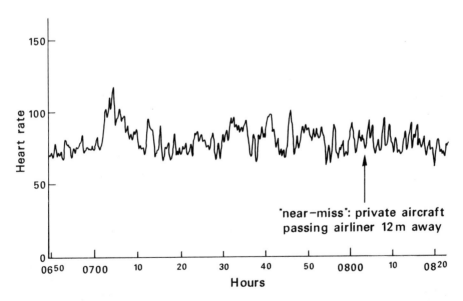

Figure 9.1. The heart rate reaction of an air traffic controller witnessing a 'near-miss' when a single engine aircraft, breaking existing regulations, passed a commercial aircraft only 12 meters away.

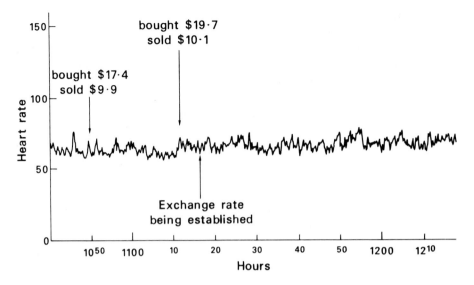

Figure 9.2. *The heart rate reaction of a foreign exchange trader, dealing with large sums of money (in millions).*

a small airfield in the north of Norway under extremely difficult weather conditions (Figures 6.8 and 6.9, Chapter 6). The pilot actually flying the aircraft during two unsuccessful attempts to land, reacted with a significant rise in heart rate (Figure 6.8), while the co-pilot in the seat next to him, and who was not involved, hardly showed any reaction at all (Figure 6.9). This may be compared with the minimal reaction of the air traffic controller in the tower witnessing the near-miss shown in Figure 9.1, when a small single-engine aircraft breaking existing regulation, passed a commercial aircraft only some 12 meters away. As opposed to this, Figure 6.7 in Chapter 6 shows the marked increase in heart rate, indicating considerable stress, in a pilot making two attempts to land at a small air strip, the first attempt being unsuccessful.

Another profession which may be said to have some stress aspects in common with the air traffic controllers is the foreign exchange trader, dealing with large sums of money, requiring considerable capacity to cope with several problems simultaneously, and the ability to make quick decisions with large amounts of money at stake. An example of such a dealer's reaction is illustrated in Figure 9.2, a reaction which appears surprisingly moderate, and not unlike the reactions observed in a number of the air traffic controllers.

Medical examination

The medical examination of the air traffic controllers, which took place in the fall of 1981, revealed, on the whole, a rather healthy group. However, nine of them had symptoms of gastrointestinal disorders, four had had peptic ulcers. Otherwise, no significant clinical findings were revealed, X-ray examination of the heart and lungs gave normal findings. Pulmonary function tests gave normal results, and ECG at rest was normal in all cases. Urine and blood examination, including blood lipids, revealed on the whole, normal findings, except in a few cases of known pathological conditions. Only ten of them were habitual smokers, and three were overweight.

Recording the blood pressure under strictly standardized conditions, showed a tendency to higher values than would be expected on the basis of age. Of the 33 subjects examined, one, who had been treated for a kidney condition, had a systolic blood pressure of 188. Five of them had systolic blood pressures over 150, 11 had 140 or higher. One subject had a diastolic blood pressure above 130, five had diastolic pressures over 100, and 14 over 90.

The blood pressure measurements were repeated in 27 of the subjects, using exactly the same standardized technique, one year later, in the fall of 1982. The mean systolic blood pressure, which was 136 mm Hg in the first study, had now dropped to 116 mm Hg. The diastolic pressure was 90–87 in 1981, as against 75–73 in 1982 (Table 9.2). The difference is statistically highly significant ($p < 0.001$). The blood pressure was lower in 1982 in all of the subjects except one. He was subsequently subject to a further examination at a hospital, at which time his blood pressure was found to be normal. The only apparent difference between the two blood pressure examinations (1981 and 1982) was the fact that the air traffic controllers were engaged in an acute labour relation conflict in 1981, a conflict which was entirely settled by the time the second blood pressure measurements were made in 1982.

It should be noted that in spite of the marked drop in mean blood pressure from 1981 to 1982, the mean resting heart rate was unchanged (Table 9.2). This may indicate that the elevated blood pressure was caused by increased peripheral resistance, rather than by increased cardiac output due to increased heart rate.

The most interesting feature of this study is undoubtedly the finding of a transitory elevation of the blood pressure, associated with the involvement in a labour relation conflict. There appears to be little doubt that the observed transitory blood pressure elevation was caused by the conflict. It is interesting to note that the same transitory blood pressure elevation also occurred in the personnel manager, representing the employer's side of the conflict. This is not the first time a clinically

Table 9.2. Resting heart rate and blood pressure in 27 air traffic controllers in 1981/1982

Subject no.	Resting heart rate		Blood pressure	
	1981	1982	1981	1982
1	66	72	140/86–84	110/70–66
2	64	63	142/86–82	116/70–68
3	58	62	128/88–86	94/64–60
4	96	80	152/108–104	168/120–118
5	71	68	124/82–78	100/68–66
6	75	56	124/80–78	108/70–66
7	60	80	138/94–90	120/76–76
9	64	59	122/100–98	108/74–74
10	51	51	132/90–88	116/70–68
11	56	64	140/82–78	110/74–74
14	71	90	154/96–92	124/76–76
15	78	79	128/90–88	110/78–78
16	67	65	128/100–96	102/70–68
17	75	69	130/84–80	104/74–70
18	58	75	132/86–82	116/70–66
19	49	61	152/104–100	126/76–76
20	60	81	128/78–74	120/80–76
21	50	44	122/80–78	104/66–60
22	53	61	130/92–90	114/76–76
23	46	43	120/82–80	106/66–66
24	49	46	140/88–86	120/80–80
26	80	82	134/84–80	120/74–74
27	72	56	154/94–90	142/86–84
49	62	62	118/80–78	102/68–64
52	57	53	128/80–76	108/68–66
53	58	56	142/94–90	110/62–58
54	70	58	188/132–130	146/94–94
$\overline{M}$	64	64	136 90 87	116 75 73
SD	11.5	12.4	14.8 11.5 11.6	15.6 11.3 12.0

Note. Resting heart rate and blood pressure (systolic and diastolic, phase 4 and 5) in the same 27 air traffic controllers in 1981, during a major labour-relation conflict, and again in 1982 when the conflict was over.

recorded elevation of blood pressure has been reported in groups exposed to similar stressful situations. Thus Rognum and Erikssen (1980) observed a significant elevation of the blood pressure of men employed by a company at a time when the company was threatened by bankruptcy. The following year, when the company's financial difficulties were overcome, the blood pressure had returned to normal. Furthermore, Anderson (1983) has shown a relationship between the stress of loneliness and elevation of blood pressure in older women. He observed a significant reduction in blood pressure when the subjects were removed from their isolation and included in a programme of regular social contact. Probably the first published suggestive evidence of any relationship between emotional stress and blood pressure was the observation by Alvarez and Stanley (1930) of a considerably higher blood pressure in the guards than in the prisoners in a prison in the USA. They thought that this might be partly due to the fact that the guards, at the time of the measurements, were applying for a position, and 'many were anxious for fear something would be found to cause their rejection'.

It is thus clear that labour relation conflicts causing prolonged emotional engagement may cause a transient elevation of the blood pressure, both systolic and diastolic. The question remains, however, how long such a situation has to be maintained for the transient elevation of the pressure to become permanent, and may be classified clinically as a state of hypertension. From the above reported studies, it is evident that it takes more than a few months. In a study still in progress, involving workers of a large ferro alloy production plant, a sudden and marked elevation of the blood pressure was noted in the workers at a time when it became likely that the plant would be closed down due to inability to operate at a profit. This threatening situation has prevailed over a period of more than three years, and the mean blood pressure level has remained elevated all these years.

This relationship between abnormal social situations, including labour relation conflicts and hypertension, should be kept in mind when considering blood pressure levels in occupational groups. It may in fact well explain the conflicting results reported in the literature concerning the blood pressure of air traffic controllers. Some of the measurements may have been made during periods of conflicts, while other measurements may have been made during periods when there were no conflicts.

On the basis of the clinical examination one is left with the impression that the air traffic controllers examined, on the whole, appear to be a fairly healthy group. It is an open question whether or not the relatively large number of cases of dyspepsia is a peculiarity of the type of people who are attracted to this profession, or due to their life-style, including irregularity of meals, excessive coffee consumption, irregularity of sleep, etc., or whether or not it is related to the work itself. On the basis of this study it is hard to see that the latter is the case. It is true that the air traffic controllers work in shifts. This may cause some irregularity in their life rhythm, especially in the case of those who live far away from the airport and have to travel long distances to work, unless they elect to spend the night prior to the morning shift at the airport.

It should also be noted that the stress associated with the labour relation conflict situation in this particular case was not a matter of being afraid of losing their jobs. It was a matter of suffering a sense of being not only unpopular, but even largely disliked by the airline passengers, and to some extent by the community at large. This underscores the fact that one should not limit any possible negative effects on health to conditions at work, but keep in mind that many external factors may also play a role in causing stress-related health hazards. This possibility should be kept in mind when considering cause and effect of stress-related complaints.

Even though it might be desirable to develop a system for air traffic control which may be easily operated by persons of average ability, it is generally agreed that at present this is not possible. The present system for control and surveillance of air traffic requires employment of individuals who are specifically selected and trained for the task. Merely to provide more radar screens, and to employ more people to use them, does not, by itself, solve the problem. The solution, in the final analysis, lies in the individual air traffic controller's ability to conceive, interpret, analyze, judge, react and to act adequately (Gerathevold, 1973).

In summary, then, it may be concluded that the findings of Mundal *et al.* (1983) support the impression gained from previous studies that the work of the air traffic controller may be demanding. Yet, our results do not indicate that the air traffic controllers are subject to unreasonable levels of stress. The state of health of those examined appears, on the whole, to be good, although the symptoms and signs of gastrointestinal complaints appear to be somewhat high. The elevated blood pressure observed in 1981, by all indications, was probably the result of a major labour relation conflict, involving the air traffic controllers in question. The conflict was solved in 1982, at which time the blood pressure had returned to normal levels.

Chapter 10
Industrial heat stress

As pointed out in Chapter 3, the ambient temperature is one of the many factors which affect performance (see Figure 3.1). The available methods for the assessment of the stress imposed by the temperature of the working environment, and its effect on the worker, were described in Chapter 5. In this chapter we shall briefly discuss some aspects of the practical application of some of these methods in heat-exposed industry.

Field studies of industrial heat stress

For a number of years the Norwegian Institute of Work Physiology in Oslo was engaged in a series of extensive field studies of heat stress in several types of heat-exposed industries, especially the electrochemical industry. This included the production of steel, ferroalloys, aluminium, magnesium, siliconcarbide, and paper, in addition to studies of sailors on board ships operating in tropical waters. In most of these studies, measurements of the environmental heat load were made, and the thermo-regulatory response of the individual worker assessed. This is a brief summary of some of the results. For a more extensive review of these studies, see Magnus *et al.* (1980), Egeland *et al.*, (1981) and Rodahl and Guthe (1988).

Two series of studies, one in the winter and one in the summer, were made at the ferroalloy plant Fiskaa Verk in Kristiansand. The purpose was to make a survey of the heat stress encountered in this type of industry, and to investigate its effect on the workers, especially in the light of the findings of Kloetzel *et al.* (1973), which suggested a connection between heat exposure and the development of hypertension in workers in the smelting industry.

The first series of studies took place in 1978, involving 12 workers. Botsball readings revealed ambient temperatures above the recommended ceilings at all examined work areas in the furnace room (26–38 °C WGT).

The heart rate was recorded continuously by a miniature portable recorder. It revealed considerable circulatory stress, not only in heat exposed workers (in one case a heart rate of 130 beats·min^{-1} for 20 per cent of the observation period), but also in manual workers engaged in the breaking up of finished metal with sledge-hammers in a cool environment (in one case a heart rate of 125 beats·min^{-1} for 45 per cent of the observation period was recorded.)

Skin and rectal temperatures were also recorded continuously. The rectal temperature exceeded 38 °C in all but two subjects. In one of the tappers the rectal temperature remained above 38 °C for 87 minutes, and in a maintenance worker for 93 minutes. Rectal temperatures in excess of 38 °C were also observed in one of the manual workers not exposed to ambient heat, indicating that the rise in rectal temperature was due to muscular work and not to ambient heat stress (Figure 10.1). Rectal temperatures in excess of 38 °C for extended periods were also observed in maintenance workers engaged in repair work by the pot, with heart rates up to 180 beats·min^{-1}.

Sweat loss averaged about 2 liters during the working shift, the highest value recorded being 3.6 liters in the case of one of the tappers. The hourly sweat rate, on the other hand, could in some cases be rather

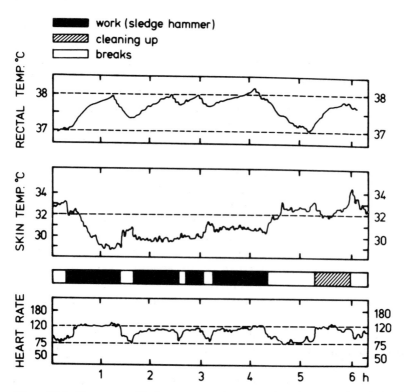

Figure 10.1. Heart rate and body temperature in a manual worker engaged in the breaking up of finished metal with sledge hammers in a cool environment in a ferroalloy plant.

high, in one case as high as $0.508 \, l \cdot h^{-1}$. It was observed that only about half of the fluid lost through sweating was replaced by fluid intake during the working shifts, leading to hypohydration. The degree of dehydration expressed in percentage of the body weight was 1.15 per cent on the average. This is close to the level at which objective signs usually occur in the form of increased heart rate and impaired endurance (Åstrand and Rodahl, 1986).

The blood studies revealed plasma renin concentrations significantly higher during heat exposure compared to values before the beginning of the working shift ($p < 0.05$). This was interpreted as an indication of reduced kidney blood flow, due to heat or physical work or both, since this elevation also occurred in the manual workers not exposed to heat but engaged in strenuous physical work. The serum concentrations of creatinine, urea and uric acid showed a significant increase in the course of the working shift, most pronounced in the case of creatinine ($p < 0.001$).

The urinary analyses showed reduced Na and K excretions during the working shift, compared to values off work. Creatinine-clearance, which is a measure of the glomerular filtration in the kidneys, was, on the average, lower during the working shift than was the case off work. The greatest difference in creatinine clearance between work and leisure was observed in the manual workers engaged in exceptionally hard physical work, who also had the highest nor-epinephrine eliminations. The mean urinary epinephrine elimination was on the whole, quite similar during the working shift and during the time off work.

The 'logical reasoning test' as an index of mental fatigue or performance administered to all the subjects during the study, before and after the working shift, and showed no difference of statistical significance.

Blood pressure, both systolic and diastolic, was measured in a standardized manner before, during and after the working shift in ten subjects, and showed no significant difference. Nor did the measurement of blood pressure in 50 heat-exposed and 50 non heat-exposed workers in this plant. Subsequent, unpublished, extensive studies of the blood pressure and the incidence of hypertension in workers of this plant, in whom the blood pressure had been followed over several years, have verified this finding: exposure to heat stress is not associated with the development of high blood pressure, as previously claimed by Kloetzel *et al.* (1973).

A subsequent series of studies at this plant (Egeland *et al.*, 1981), verified or substantiated the findings of the first study. In addition, it emphasized the importance of acclimatization and indicated the occurrence of some degree of hemodilution during the working shift in the heat exposed workers. It pointed out specifically that no ill effects were observed as a consequence of the heat exposure encountered at this plant, and that all the physiological parameters measured were within the normal range. It appeared that one person was susceptible to ortostatic hypotension, and was apt to faint when heat exposed, and was thus unsuited for this type of work. This also applies to persons suffering from cardiovascular disease or chronic kidney disease. Otherwise, no evidence of health hazards were detected for these kinds of industrial operations. The seasonal differences in the observations were minimal.

Extensive studies of Norwegian aluminium plant workers at the Lista Aluminium Plant, using the technique described above as well as fluid balance and Botsball temperature recordings, revealed ambient temperatures considerably above the recommended upper limits at a number of job locations, (Rodahl, 1981; Jansen *et al.* 1982). The studies were made twice a year for two years. There was suprisingly little difference between winter and summer temperature in the plant, i.e. only about 10 °C.

The workers in this particular aluminium plant were exposed to the heat of the pot-room for 44 per cent of the total shift, on the average. The rest of the time was spent in the canteen drinking coffee, talking, reading, playing cards or moving about. As is often the case, the workers preferred to work rather intensely in order to get the job done as quickly as possible, to be able to rest that much longer. The result may be extended periods of excessive heat stress instead of more frequent brief periods of exposure interspersed with brief cooling-off periods outside the pot-room. A burner cleaner may serve as an example (Figure 10.2). The time spent actually cleaning the burners amounted to three one-hour periods. The rest of the time was spent outside the pot-room in a comfortable cool canteen. The time spent by the pot, on the whole, varied from 30 to 130 minutes per shift.

In 15 out of 22 workers the rectal temperature exceeded 38 °C. An example is given in Figure 10.3. In one case the rectal temperature exceeded 38 °C for 63 per cent of the observation period. When exposed to the pot-room temperature the subject's rectal temperature rose gradually, levelling off in 45–60 minutes, whereas the skin temperature reacted much faster and is a good indicator of how close the subject is to the heat source. In our experience, skin temperatures over 36 °C are usually associated with profuse sweating, which is also in accordance with experimental data (Nielsen, 1969).

In almost all subjects studied, the heart rate exceeded 110 beats $\cdot$ min^{-1} for considerable periods of time, in some cases up to 30 per cent of the time. In several cases it exceeded 130 beats $\cdot$ min^{-1}, which in most people represents about half of the heart rate reserve (halfway between resting and maximal heart rate).

In several cases the weight loss of the workers exceeded one per cent of the body weight in the course of the shift, indicating a considerable

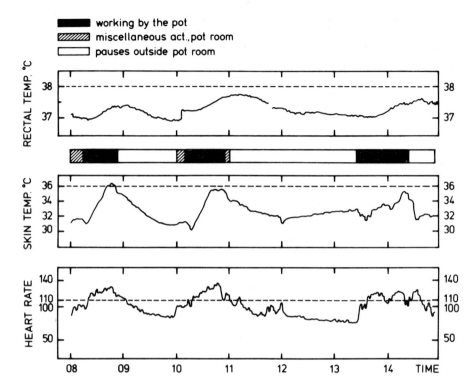

Figure 10.2. *Work-rest pattern in a burner-cleaner in an aluminium plant.*

negative fluid balance, or hypohydration. The highest sweat rates which we observed were about $0.8\,l\cdot h^{-1}$. One worker consumed as much as 6 liters of water in the course of his working shift. This excessive fluid loss due to sweating is a major problem in workers regularly working in our heat-exposed industries.

Our blood studies showed a lower hemoglobin value at the end of the shift compared to the values before the shift (15.4 ± 0.2, as against 14.7 ± 0.3), indicating hemodilution. Similar findings were made in our studies in the ferroalloy industry. We also found that those who had the greatest sweat losses also had the greatest drop in sodium concentrations of the blood, but in no case did we find values which were outside the normal range. In every case the sodium loss was replaced by the time the subject appeared on the shift the next day. We therefore concluded that there was no advantage in taking salt tablets during work.

Urinary catecholamine elimination was also examined in our studies, but was found to be of no value as an indication of stress under these conditions. Individual differences were greater than fluctuations with stress exposure in the same individual.

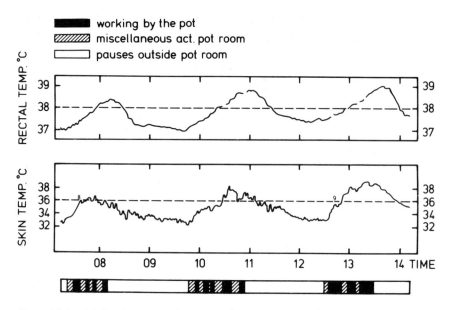

Figure 10.3. *A jack raiser in an aluminium plant; an example of rectal temperature exceeding* 38°C.

From this brief review it appears that with the ambulatory loggers now available, it is possible to move the laboratory out into the field, and to carry our rather sophisticated scientific studies on ordinary workers during the actual performance of their work at their work places, with the same degree of sophistication as in the laboratory.

Safe limits of heat exposure

It is generally agreed that there is no single index or parameter which completely reflects the strain to which a worker is exposed during his work in heat-exposed industry. It appears that too much emphasis has been placed on the academic accuracy at the expense of practicability. This is probably due to lack of practical experience as to the actual prevailing conditions in heat-exposed industry on the part of those who have developed the indices, which are essentially based on laboratory experiments. This may have caused them to overlook the fact that work in these cases on the whole, is performed under rather varied conditions, with varying degrees of physical work load, heat stress and work periods. Industrial work of the kind in question is very seldom comparable with the schedules used in laboratory experiments. For this reason, the suggested so-called safe upper limits of heat exposure under conditions of continuous work, and work 75, 50, or 25 per cent of the time, may be artificial and therefore not applicable in real-life situations.

Furthermore, the workers move frequently from place to place, with greatly varying ambient temperatures. In addition, the thermal stress varies with time, from hour to hour, and from night to day (see Figure 10.4). Finally, different members of the same work team may be exposed differently even though they are performing the same type of work. Differences in the type of clothing used may also affect the workers differently, even though they are engaged in the same work operation. Individual differences, and degree of acclimatization, must also be taken into account, as well as the level of fitness in relation to the physical work load to be performed. In addition, age and sex have to be taken into consideration. In our studies in an aluminium plant, the amount of time each worker spent in the pot-room amounted to some 44 per cent of the total period of the shift. The rest of the time was spent outside the pot-room, at ordinary room temperature. Of the 44 per cent of time spent in the pot-room, 34 per cent was spent working. Similar findings were made in different ferroalloy industries.

Permissible upper limits for safe heat exposure have been based on the assumption that the rectal temperature of the worker should not

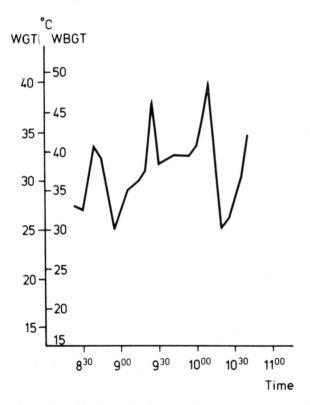

Figure 10.4. The fluctuation in the ambient temperature in one location in the pot-room in an aluminium plant.

exceed 38 °C. At ordinary room temperature this is usually the temperature reached in continuous-cycle ergometer work in the laboratory at a work rate representing approximately 50 per cent of the individual's maximal oxygen uptake (I. Åstrand *et al.*, 1975). But when working in the heat, part of the body temperature rise is due to the heat gain from the environment, which is added to the internal heat mainly produced by the working muscles. As is evident from our field observations, core temperatures exceeding 38 °C are rather frequent, without this appearing to have any harmful effect. In fact, one of our subjects came to work on a bicycle and had a rectal temperature of about 38 °C upon arrival, due to the work of bicycling. The same applies to the sophisticated heat indices, which hardly provide a better assessment of the stress of heat than that felt subjectively by the worker himself. This does not mean that a reliable thermal index is of no interest. There is certainly a need for an objective assessment of the thermal environment, especially in order to assess objectively whether or not any of the changes made, in fact, represent an improvement of the environment.

The heat stress may be more or less uniform throughout the work shift, as may be the case in a pot-room. Or it may fluctuate markedly with brief periods of extensive heat exposure, as may be the case in a ferroalloy plant during tapping, or in a paper mill during certain maintenance procedures. In the latter cases, mean values for heat exposure may be questionable. Prolonged moderately high heat exposures may be equally as bad or even worse as brief intense heat exposures. In addition, humidity may be a decisive factor for tolerance. For a critical review of the merits of some of the most commonly used indices, see Rodahl and Guthe (1988).

Health hazards of industrial heat exposure

Both occasional and regular heat exposure may represent health hazards. Best known are the ill effects of acute exposure, i.e. fainting, heat exhaustion and heat stroke. How frequently these disorders occur is not exactly known, due to inconsistency of diagnostic criteria and reporting routines. For references, see Knochel *et al.* (1961), Khogali and Hales (1983) and Dinman and Horvath (1984).

The different ill effects which have been claimed to occur in individuals occupationally exposed to heat stress over prolonged periods, are less well established (Dukes-Dobos, 1981). In 1973, Kloetzel *et al.* reported their finding of a relationship between hypertension and prolonged exposure to heat. One of the authors had noticed an unusually high prevalence of hypertension in workers exposed to high heat levels, particularly furnace workers, in a small steel plant in

Brazil. Later on they pursued this observation in more detail in another steel plant in Brazil. In a preliminary survey they examined blood pressure readings obtained during the past years in a group of selected workers, on the basis of records from the files. Their figures showed relatively more cases of blood pressures over 140/90 in heat exposed workers than in clerical staff employees, apparently supporting the initial impression. The subsequent study, the results of which form the basis for the quoted publication of Kloetzel *et al.* (1973), involves a total of 330 individuals, of whom 90 were exposed to no heat. It should be noted that the median age of the 'no-heat' exposed individuals was 25.5, as against 36.8 years in the 'extreme-heat' exposed ones. It should also be noted that the blood pressure readings were made only once in each subject, at the end of a 30 minute rest, in connection with routine medical examinations of the workers. The salt intake was not recorded, but the authors' claim that the salt dispensers were avoided by the workers because of rumours that salt tablets may cause impotence! They found a statistically significantly higher incidence of blood pressure over 140/90, and over 160/95 in the heat exposed group compared with the group not exposed to heat. They also observed an increasing incidence of hypertension with the number of years the individuals had been exposed. The authors indicated further studies in an attempt to identify the causes of hypertension in individuals exposed to high levels of heat. As far as one can see, no such studies have as yet been published.

Lund-Larsen and Dahlberg (1981) were unable to confirm the observation of Kloetzel *et al.* (1973) in their study of 551 men at an aluminium plant in Norway (Årdal-Sundal verk) in 1975–76; nor did we in a study of the workers at the ferroalloy plant Fiskaa Verk, in 1978 (Magnus *et al.*, 1980). There was, as already mentioned, no statistically significant difference in the blood pressure in 50 heat-exposed workers and 50 workers who were not exposed to heat. This finding has been confirmed in a more extensive longitudinal study in the same plant, still in progress, by a team headed by Erikssen *et al.* (1988, unpublished). In the study of Magnus *et al.* (1980), a significantly increased plasma renin activity was observed in heat-exposed workers, but this was also the case in other workers at the same plant who were engaged in hard physical work in a cool environment. The elevated plasma renin activity observed in this study was therefore interpreted as being an expression of reduced kidney blood flow due to intense heat or due to strenuous physical work, both of which are known to cause reduced kidney blood flow (see Åstrand and Rodahl, 1986). A drop in kidney blood flow is observed during heat exposure at rest, and more so if the heat-exposed individual is subject to physical activity (Radigan and Robinson, 1949). At maximal work rates in the heat, a 70 per cent drop in renal blood flow has been observed (Rowell, 1974). Kosunen *et al.*

(1976) observed a two-fold increase in renin concentration after 20 minutes in a Finnish Sauna bath at a temperature of 85–90 °C; the increase was significant even after 10 minutes. Finberg and Berlyne (1977) have shown that the increase in renin is less in heat acclimatized than in non-acclimatized individuals. An increase in renin concentration was observed in physically active subjects at normal room temperature, at work rates corresponding to 70 per cent max VO_2 (Kotchen et al., 1971). It thus appears that although it is generally accepted that elevated renin production over long periods of time may lead to hypertension, the elevated renin levels observed in our heat-exposed subjects are not associated with hypertension.

In the course of the mentioned blood pressure study, in workers at the ferroalloy plant, headed by Erikssen (1988), a sudden elevation of blood pressure occurred in a number of the individuals being studied, during the fourth year of the study, at a time when a closing down of the plant was being considered due to a serious deficiency in the earning capacity. This blood pressure elevation, evidently, was an expression of an autonomous nervous system response to stress, and not due to heat exposure (see Chapter 9).

Resmond et al. (1978) made an extensive study of the mortality of workers in different heat stress categories in 59,000 steelworkers. They observed that workers in jobs involving higher environmental heat showed less risk of death from cardiovascular disease, the risk decreasing with increasing length of exposure. Work load did not appear to be an independent contributory factor. Workers with less than six months exposure showed higher risks of death from cardiovascular diseases. They also observed an increased risk of mortality from digestive disease from workers employed in jobs in high heat categories. Pavlenko (1971), on the other hand, found that the incidence of circulatory diseases was significantly higher among workers in hot shops than among those working in cold shops in the metallurgical industry.

In a study of 1,000 workers at 20 different foundries, Hernberg et al. (1976) found that the systolic and diastolic blood pressure of carbon monoxide-exposed workers were slightly higher than those of other workers, when age and smoking habits were taken into consideration. However, exposure to heat could not be separated from exposure to CO in this study.

Gusic et al. (1969) have reported clinical and histological changes in the mucosa of the nose in workers in an electric light bulb factory who were exposed to heat.

An association between heat exposure and cataracts has been claimed since the early eighteenth century. Wallace et al. (1971) examined two groups of steel workers, who differed widely in their heat exposure. Although a higher prevalence of the common form of cataract was found in the heat-exposed group than in the non-exposed group, only two cases of cataracts were found that would be generally accepted as occupational in origin. David and Popescu (1972) detected a significant degree of accommodation insufficiency in heat-exposed men working in the forge section of an iron and steel mill. Jebava *et al.* (1980) reported a high incidence of cataracts in glassblowers, three quarters of whom neglected the wearing of protective glasses. A decreasing trend in the occurrence of glassblowers' cataract was observed due to the improved glass technology in Czechoslovakia.

Cutaneous implications of excessive heat in the place of work have been described by Olumide *et al.* (1983), and by Dibeneditto and Worobec (1985).

Heat exposure has been claimed to reduce fertility due to impairment of semen quality and sperm abnormalities. For references, see Baird (1985). According to Macleod and Hotchkiss (1941) a body temperature rise to more than 40 $^{\circ}$C for a short period of time (1–2 hours) induces a reduction in the sperm count which may persist for some 2–3 months. Elevated body temperatures for prolonged periods of time, i.e. several days, as in the case of typhoid fever or viral diseases, may in some cases cause a transient suppression of spermatogenesis (Macleod, 1967). While the adverse effect of testicular hyperthermia is well known, the question remains as to whether or not the changes described in the case of hyperthermia are in fact caused by the elevated temperature as such or by the infectious agents and their toxins underlying the fever of the patient.

Application of existing knowledge

The safeguarding of workers employed in hot industries requires an understanding of the physiological reactions to heat stress, and more so when industrial heat exposure is superimposed on a tropical climate. It is also essential to distinguish between dry heat often prevailing in Nordic industrial environments, and humid heat encountered in many other parts of the world, especially in tropical or semitropical areas. To cope with heat stress may therefore to a considerable extent be a matter of education, knowing how to behave, how to space work and rest, and how to replace sweat loss by fluid intake. In this respect, the US Department of Labor and the US Department of Health and Human Services' pamphlet *Hot Environments* (1980) is a good example.

Fernandez (1980) has described a device called the Thermoguide, i.e. a card containing information concerning the hazards of excessive heat exposure, what to do and what to expect. Zal (1984) has prepared a recommended programme in an easily understandable language for employees exposed to extremes of heat. Similar material, including an audiovisual programme, has been prepared by us for use in Norwegian aluminium and ferroalloy smelters.

In laboratory experiments, Kamon (1979) studied different exercise cycles in six heat-acclimatized male subjects exercising at 40 per cent VO_2 max in hot ambient conditions. Based on the leveling off of heart rate and rectal temperature, and setting the limit of a rectal temperature at 38 °C, he found that a schedule of exercise and rest periods of 20 minutes was adequate for a hot-dry ambient condition, with a dry bulb temperature of 50 °C, wet bulb 25 °C. Essentially similar observations were made by us in real life in a Norwegian magnesium plant at a somewhat lower work load with a work schedule of 20 min work interspaced with a 10 minutes cooling-off period at prevailing outside ambient temperatures (see Figures 10.5 and 10.6).

Considerable efforts have been devoted to the development of cooled garments, suitable for use in hot industrial work places. A number of studies have been made to evaluate the efficiency of such garments (Shvartz *et al.* 1974; Mairiaux *et al.*, 1977; Hausman and Petit, 1979; Pasternack, 1978; Engel *et al.*, 1984). This includes the value of reflecting layers in clothing (Kerslake, 1969). Goldman (1974) has reviewed the problems involved in clothing design for comfort and work performance in extreme thermal environments. Thomas (1974)

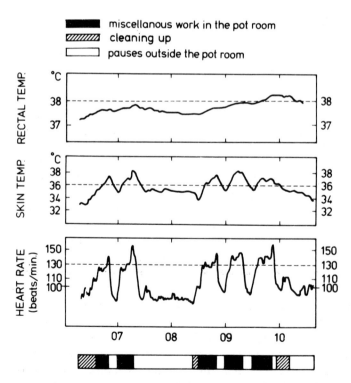

Figure 10.5. Worker in a magnesium plant, slag removal; original work schedule.

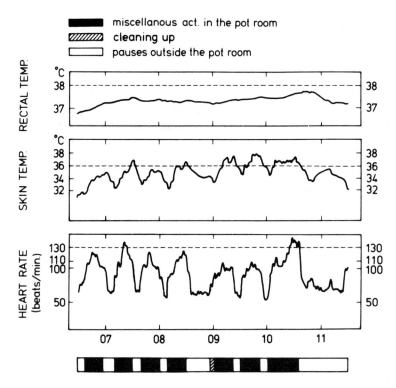

miscellanous act. in the pot room
cleaning up
pauses outside the pot room

Figure 10.6. *Worker in a magnesium plant, slag removal; modified work schedule (20 min work, 10 min cooling-off break).*

showed that simple heart rate measurements were useful in the evaluation of the efficiency of compressed air for heat-exposed maintenance workers. Martin and Callaway (1974) have shown that the wearing of a protective face mask imposes a significant additional heat stress, manifested by increased sweat rate, increased rectal temperature, increased skin temperature and increased heart rate. Shvartz *et al.* (1974) studied the effect of cooling ten different body regions by circulating water, and concluded that for an efficient design of a whole body cooling suit, the head and upper extremities should receive 15 per cent of the total amount of water tubing used, the torso 38 per cent, and the lower extremities 32 per cent.

It appears well established that cooling of the skin of the head affects the thermoregulatory response in general, including a decreased sweating over the entire body. Evidently this is caused by a change in the hypothalamic temperature (Baker, 1982). Morales and Konz (1968) showed that in subjects exposed to heat, the use of a water-cooled hood caused the head temperature as well as the core and skin temperatures to be lower. It caused reduced sweating and permitted longer exposure to the stress environment. Furthermore, Konz and Gupta (1969) showed that localized cooling of the head during heat exposure in the form of a cooled hood caused less decline in mental performance, the sweat rate was reduced, and the increase in heart rate was less.

The effects of cooling the head were essentially confirmed by Shvartz (1970, 1976), Nunneley *et al.* (1970), and by Greenleaf *et al.* (1980). Brown and Williams (1982) showed in climatic chamber experiments that head cooling prevented an increase in both auditory canal- and oesophageal temperatures during heat exposure. Riggs *et al.* (1981) observed that facial cooling in exercising subjects resulted in a significant lowering of the heart rate, whereas no difference was detected in blood pressure or rectal temperature.

Evidently, brain function appears to be especially vulnerable to heat (Baker, 1982). It is therefore of particular interest to note that a rather unique, selected cooling of the brain is possible due to a special vascular arrangement in the head (Cabanac, 1986). During strenuous physical work or exercise, the heat production increases proportionally to the energy output, causing an increase in body temperature, most marked in the muscle. However, the temperature inside the head, i.e. in the brain, does not rise as high as in the rest of the body (Figure 10.7). Evidently, this is the result of a special arrangement in the vascular system in the head that results in a selective cooling of the brain. As pointed out by Cabanac (1986), innumerable anastomoses connect the rich subcutaneous venous plexus of the cephalic head with the intracranial sinuses. The most obvious ones are vena opthalmica draining blood from the forehead and upper face, vena emissaria mastoidea draining blood from the lower temporal vein and external ear, and vena emmisaria parietalis draining blood from the scalp on the back and crown of the head. The direction of the flow in these veins has been examined with the aid of Doppler flow probes. During hyperthermia, cooled blood flows inwards, from skin to brain. During hypothermia the flow is in the opposite direction, or there may be no flow at all. When the hyperthermia is caused by prolonged exercise such as running, the face is exposed to increased convection, which augments the cooling by the evaporation of the sweat. Bald individuals may have the additional advantage of heat loss due to sweating from the hairless scalp, since sweating from a bald scalp can be as great or greater than that of the forehead (for references, see Cabanac, 1986).

The solution of the problems associated with industrial heat exposure may be based on a combination of practical measures such as: shielding of heat source; reducing each period of heat exposure to about 20 min, interspersed with brief 10 min cooling-off breaks; elimination of strenuous physical work close to the heat source; introducing mechanical aids to take the place of manual labour wherever possible; and nursing the industrial process properly at all times so as to avoid

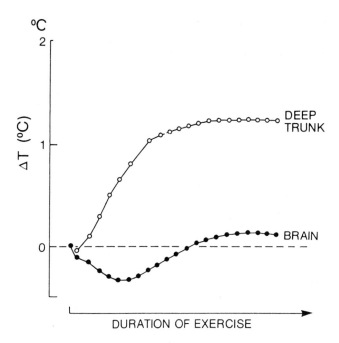

Figure 10.7. Time course of changes in deep trunk and deep intracranial temperatures during exercise in human subjects (modified from Cabanac, 1986).

complications that will necessitate drastic measures causing excessive exposure to heat stress. A key point is to avoid prolonged intense heat exposure, which will lead to profuse sweating. This avoidance of sweating will reduce the fluid loss, and thereby reduce the need for fluid intake, correspondingly. At any rate, the fluid loss should be replaced as it is being lost, and the aim should be for the worker to leave his place of work fully hydrated, able to enjoy his leisure time.

Existing plant health personnel should be supplemented with qualified work physiologists, with the necessary medical background, to make specific on-site measurements and observations in the industry. The technical and administrative managers should, in the course of their training, be given sufficient knowledge of the general principles of occupational health and physiology to enable them to use the results obtained by work physiologists to improve the working conditions, for better productivity, well-being and motivation, in the interests of both labour and management.

Chapter 11
Working in the cold

Physiological adjustments to cold stress

The ability to survive in the cold depends, in the final analysis, on the ability to maintain thermal balance. In order for the body heat content to remain constant, heat production must equal heat loss. The heat production may increase to a level about twice that at rest by shivering. A similar increase, or even greater, may take place in a patient during a fever. Heat production may increase by as much as 30 per cent following the ingestion of a large amount of meat, due to the specific dynamic action (SDA) of proteins. Physical activity, however, may increase the heat production by a factor of as much as 50. During prolonged extreme cold, physical activity is the only possible way of keeping warm when exposed to severe cold without adequate protection in terms of shelter and insulation. Under such circumstances the one who is the fittest and is able to keep up the physical activity has the greatest chance of survival, i.e. 'the survival of the fittest', literally speaking. (For references, see Åstrand and Rodahl, 1986.)

A nude, resting, individual is usually in thermal balance at a room temperature of about 30 °C. The rectal temperature is then, generally, about 37 °C, and the mean skin temperature is roughly 33 °C. If the room temperature is lowered, the exposed body will begin to cool, since the rate of heat loss to the cooler surroundings is increased. The exposed body will respond to this challenge by:

1. Vasoconstriction in the skin, causing the surface temperature of the peripheral parts of the body to drop. This reduces the heat loss to the surroundings since the temperature gradient is reduced. Such vasoconstriction may cause a six-fold increase in the insulating capacity of the skin and subcutaneous tissues (Burton, 1963). In addition the enormous variability of the blood flow through the fingers may be applied to vary the degree of heat loss through the hands (Robinson, 1963). In this manner, the finger or the toe temperature may rapidly approach the temperature of the surrounding air, an example of which is given in Figure 11.1. If the cold exposure is sufficiently

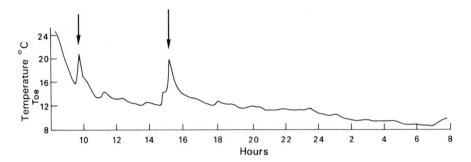

Figure 11.1. Toe temperature in a nude subject exposed to 8 °C continuously for 3 days. By the end of 24 hours, the skin temperature of the big toe had dropped to about 8 °C, the temperature of the ambient air. Arrows indicate cycle ergometer exercise.

prolonged this may cause ischemic cold injury of the exposed extremities. In this connection it is of interest to note that the blood vessels to the head are less affected by vasoconstriction than are the hands and feet. In this manner the head is, to some extent, protected against ischemic cold injury.

2. If the vasoconstriction of the skin is insufficient to maintain the thermal balance, the body has no choice but to increase the heat production, first of all by shivering. Since the mechanical efficiency of shivering is zero, all the liberated energy is transferred to heat. But this may at best only increase the heat production by a factor of about two, three at the most. The possibility of a non-shivering thermogenesis is still subject to debate (see Åstrand and Rodahl, 1986), but plays, at any rate, hardly any role of practical importance compared to physical activity, which, as already mentioned, may increase the metabolic heat production by a factor of 30–50.

Prolonged cold exposure will, as a rule, cause some adjustment to the cold environment of a psychological nature. There may also be signs of a local acclimatization, especially in the form of a higher skin temperature of the hands. The warm hands of the Eskimos are well known (Rodahl, 1953). Increased skin temperature of the hands may be achieved by keeping them in ice water for about 15–30 min daily for a couple of weeks. This local acclimatization to cold appears not to be related to any racial or ethnic differences. For further details, see Åstrand and Rodahl (1986).

While the existence of a local acclimatization to cold, then, is fairly well established, the question of general physiological acclimatization to cold in man is still an open question. A great deal of the original research forming the basis for suggesting the existence of a general physiological acclimatization to cold was done on rats. In some experiments, rats were kept in refrigerators at a temperature of 5 °C for several weeks. Those who survived had doubled their food intake, and the metabolic rate increased to twice the normal level. This was accompanied by an increase in the thyroid function. The amount of physical activity also increased markedly, and, as a rule, it was only those rats that were able to maintain a high level of physical activity that survived. However, in a number of the rats, parts of their tails became necrotic, causing bits to drop off, presumably due to ischemic cold injury on account of the prolonged persistent vasoconstriction of the blood vessels of the tail. This indicates that the state of these rats was one of pathology and not a matter of physiological acclimatization to cold.

In humans, who are natives of the Arctic, and who are protected by clothing and heated dwellings, such as the Eskimos, one does not find any of the changes which were observed in the rats kept in the refrigerator. The studies which we performed among different groups of Eskimos in Alaska showed that their food intake was comparable with that of ours. It is true that their basal metabolic rate was slightly higher than that of our white controls, but most of this could be attributed to their apprehension due to the unaccustomed situation of the study, in addition to the specific dynamic action (SDA) of their high meat diet. We also observed that the thyroid function of ordinary coastal Eskimos was no different from that of our white controls living in Fairbanks. Measurements of the temperature underneath their clothing and inside their houses indicated that they were, in fact, surrounded by a temperature environment not much different from ours. From this we concluded that humans who live in the Arctic do not exhibit significant evidence of a general acclimatization to cold. (For references, see Åstrand and Rodahl, 1986.)

On the basis of a series of extensive studies extending over a number of years, Hong *et al.* (1987) concluded that alteration in the thermoregulatory response to cold did occur in traditional Korean women divers who were subjected daily to severe cold stress. This included elevated basal metabolic rate in the cold season, and increased overall insulation of the non-fatty shell of the body. These alterations gradually disappeared in the course of three years after the divers started to wear wet suits. It should be borne in mind, however, that this type of cold exposure is quite different from exposure to cold air, and that the alterations observed in the divers may not necessarily occur in ordinary individuals exposed to cold ambient air. When nude individuals, on the other hand, are exposed to cold stress that is more like that of the refrigerated rats, and far more severe than that normally encountered by the clothed natives living in the Arctic, certain physiological changes do occur, which may be interpreted as hormonally induced responses to cold (Rodahl *et al.* 1962).

At the Division of Research, Lankenau Hospital in Philadelphia, Pennsylvania, subjects volunteered to remain continuously, for up to nine days, in a climatic chamber set at 8 °C, dressed only in socks and short underwear. This cold stress is quite similar to the one to which the rats were exposed (furred rats at 5 °C). During the nights the subjects slept under the cover of a single blanket. The rest of the time they spent resting in a chair, apart from two daily activity periods on a cycle ergometer. The experiment was repeated with four different diets; 3,000 kcal and 70 g protein; 3,000 kcal and 4 g protein; 1,500 kcal and 70 g protein, and 1,500 kcal and 4 g protein. The oxygen uptake and the heart rate were measured both during rest and maximal, as well as submaximal work loads. In addition, we measured the nitrogen balance and the urinary catecholamine elimination.

In the subjects living on an adequate diet (3,000 kcal, 70 g protein), no reduction in the physical work capacity, as judged by the maximal oxygen uptake, was observed at the end of nine days and nights at 8 °C. Cold stress combined with nutritional deprivation caused, on the other hand, a marked reduction in the physical work capacity, and the experiments had to be discontinued at the end of three days due to the development of slight ischemic cold injury of the subject's feet (i.e. ischemic cold injury of the rats tails).

The oxygen uptake at rest was higher than at comfortable room temperature. This was attributed to the shivering, since the oxygen uptake at a submaximal work load, when the subjects had warmed up and had ceased shivering, was about the same as at the same work load at room temperature. The heart rate, on the other hand, was higher both at rest and during a submaximal work rate in the cold, compared to control experiments at normal room temperature. This is probably due to increased catecholamine excretion, as a result of the cold stress.

We also observed a negative nitrogen balance during the cold exposure. This we interpreted as being the effect of an elevated thyroid function due to the cold stress.

The subjects responded to the severe cold stress by violent shivering which lasted more or less continuously, night and day, throughout the experiment, even at night when they slept under a blanket. They soon learned to continue to sleep in spite of the shivering. As a result, the increased heat production, due to the shivering, was maintained, even during sleep, so that the body temperature remained normal and the subjects slept relatively comfortably. Occasionally, however, the same subjects, for some reason, failed to keep up the vigorous shivering during the night while they slept. Consequently, their body temperature continued to drop and approached dangerously low levels, and the subjects had to be woken up. Under these conditions the subjects sometimes objected to being disturbed, and wanted to be left in peace to continue to sleep. If this were allowed the subjects might have continued to cool, and might conceivably have died from hypothermia. It thus appears that an exposed person sleeping in the cold may actually freeze to death if the body's rate of cooling is slow and gradual, so that shivering is not elicited.

From these observations, and other available data, we concluded that if general acclimatization to cold, in fact, is possible, it is of limited practical importance, compared to the importance of psychological adjustment, proper equipment, protection, experience, and knowledge of how to behave and what to do.

Arctic coal mining

Even in sheltered work places, such as in Arctic coal mines, the cold may present a problem for the workers, as is the case in the Spitsbergen coal mines. Here, contrary to coal mines in the rest of the world, the temperature is quite low in the mines due to the permafrost. Because of the geological conditions, with almost horizontal sedimentary layers, and a thickness of the coal seams from 70 to 110 cm only, the miners have to work lying on the ground. Holes are drilled along the coal face, charged with explosives and the coal deposits are blasted free from the seam. The coal is then pulled out to the transport system by means of rectangular scrapers. The roof along the new face is secured by a row of supports. The narrow passages, and the minimal height at the working sites along the coal face, forces the workers to creep or crawl several hundred metres. The work is performed in lying, half-sitting or squatting positions for two sessions of approximately three hours each, in each shift period. The permafrost causes a nearly constant temperature in the mine all the year round, 2–4 °C below zero. This seems to be accepted by the workers as a rather comfortable temperature for heavy physical labour, but it has always been regarded as difficult to keep the feet warm in this sitting or lying position.

The uniqueness of the Longyearbyen coal mines, as far as the miners are concerned, is the awkward working position, necessitated by the narrow coal seam, the low temperature in the mine, and the isolation of the mining community.

The work stress of coal miners generally has been the subject of a number of studies (Ross, 1972; Chakraborty *et al.*, 1974; Althouse and Murrell, 1977; Oleinikov, 1975; Vaida and Pafnote, 1979; Hieber and Smidt, 1982; Bobo *et al.*, 1983). No studies, apparently, have been made of the work stress of coal miners working under conditions such as those encountered at the Longyearbyen mines. A study was therefore undertaken to assess the work stress of coal miners operating in low coal seams in very restricted surroundings under Arctic conditions (Alm and Rodahl, 1979).

Four miners, 21–29 years of age, engaged in actual mining operations in one of the typical low-seam mines at Longyearbyen, volunteered to serve as subjects for the study. Their body size and estimated
physical work capacity are listed in Table 11.1. They were studied for
24-hour periods, both during work in the mine, and during time off
and sleep. The study included assessment of maximal oxygen uptake
based on the recording of heart rate during submaximal cycle
ergometer exercise (Åstrand and Rodahl, 1986); assessment of physical
work load based on the continuous recording of heart rate with the aid
of shielded miniature portable magnetic tape recorders (Oxford Medilog); assessment of thermal stress based on continuous recording of
rectal and skin temperature (thigh) by the same Medilog recorder; and
the assessment of general stress response on the basis of the analysis of
urinary catecholamine eliminations (Andersson *et al.*, 1974).

The estimated work load for the four subjects ranged, on the average,
from 1.1 - 1.6 l $O_2 \cdot min^{-1}$, corresponding to about 30–40 per cent of
their maximal oxygen uptake (Table 11.2). Adrenaline excretions
reached levels up to 25 nanograms $\cdot min^{-1}$ during work (Figure 11.2).
The rectal temperature ranged from 37.5 to 38.5 during work. In spite
of this high rectal temperature, the skin temperature of the thigh fell in

Table 11.1. *Age, body height and weight, and physical work capacity of four miners.*

Subject no.	Age years	Height cm	Weight kg	Physical Work Capacity	
				Max O_2 l·min^{-1}	ml·kg^{-1}
1	23	199	91	2.9	32
2	21	177	78	4.2	54
3	29	182	91	3.5	38
4	26	182	85	4.1	48

Table 11.2. *Estimated work load*

Subject	Work load		
	Mean work pulse	Estimated mean O_2 uptake l·min^{-1}	Estimated % of VO_2 max (approx.)
1	120 (100–150)	1.2	40
2	110 (80–140)	1.5	35
3	105 (85–130)	1.1	30
4	105 (80–135)	1.6	40

URINARY CATECHOLAMINES

COAL MINERS, LONGYEARBYEN

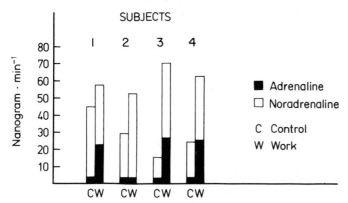

Figure 11.2. *Mean urinary catecholamine elimination (adrenaline and noradrenaline) in the four Spitsbergen coal miners during work, compared with night values.*

SUBJECT NO. 1 SVALBARD 14 MARCH 1979

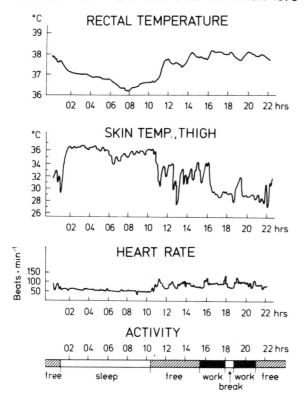

Figure 11.3. *Body temperature (skin and rectal) and heart rate in subject no. 1 (coal miner) in Spitsbergen.*

two of the subjects to about 28 °C (Figure 11.3). In the two remaining subjects, the skin temperature remained around 33 °C.

An example of a detailed analysis of the recorded heart rate of subject no. 1, paired with the activity record, is presented in Figures 11.4 (off duty) and 11.5 (on afternoon shift). This example is, in general, quite typical for all four subjects, and may therefore be considered representative. On the whole, the level of physical strain, in terms of heart rate response, is about the same during leisure activity as during work in the mine. In the example shown in Figure 11.5, the subject works about 5.5 hours in the mine, interrupted by a half-hour break. During work he maintained a fairly steady work pace for extended periods, especially during the first half of the work shift. It is evident from the record that this type of mining operation may impose some rather unique types of stress on the miner, as is the case when, at the start of work, he crawls along the narrow passage, pulling a box containing 50 kg dynamite tied to his leg, causing his heart rate to approach 165 beats · min^{-1}.

Although this study only involved four subjects, the uniformity in their response indicates, on the whole, that the results may be considered fairly representative for this type of mining operation. Compared to other occupations in which the same methods of assessment have been used, it is reasonable to conclude that the miners operating these low-seam coal mines in Spitsbergen are subject to considerable

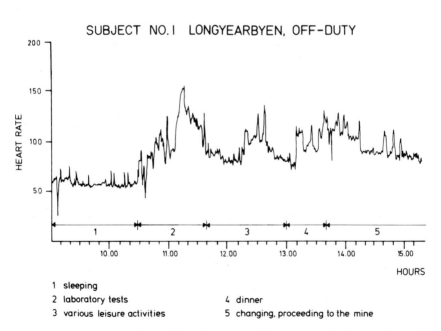

Figure 11.4. *Heart rate in one of the coal miners (subject no. 1) in Spitsbergen during off-duty activities.*

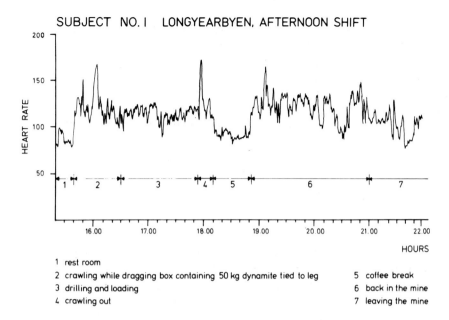

Figure 11.5. Heart rate in the same coal miner shown in Figure 11.4, during afternoon shift.

physical work stress. Physical work loads of this magnitude (30–40 per cent of VO_2 max) exceed work loads commonly encountered in most industries, where the work load seldom exceeds 25 per cent of VO_2 max. The work load of these coal miners is comparable to that of coastal fishermen, in whom the levels of urinary catecholamine excretion equalled those observed in the present coal miners. It also appears that at least some of these coal miners may be subject to some degree of cold stress of the lower extremities while operating the low-seam mines, necessitating crawling and working in a stooped or lying position in an ambient temperature slightly below freezing.

Protective clothing

American studies have indicated that the slightly elevated metabolic rate during work in severe cold was not caused by the cold, but by the hobbling effect of the Arctic clothing (for references, see Consolazio, 1963; Rodahl, 1963). It is therefore important that working garments, to be used under severe cold weather conditions, should not only have a high insulating value, but also provide mobility. This, in the case of the Eskimo, is achieved by a few layers of fur, providing all the insulation needed, while at the same time being loosely fitted so as not to restrict the full range of mobility of the extremities.

A graphic presentation of the approximate insulating requirement at different activities and temperatures is presented in Chapter 5, Figure 5.7. A schematic indication of the insulating value of different kinds of natural furs is given in Figure 5.8. Since it would not be possible, at best, to provide everyone who needs such protection with clothing of natural fur, it is indeed encouraging to know that, with the technical resources now at our disposal it is possible to produce fabric made of artificial fiber, which may be equally good, if not better, than natural fiber, and which, in contrast to natural fiber, may be modified as needed to meet a variety of special requirements. An example of a combination of fabric made of artificial fiber is presented in Figure 11.6, simulating, to some extent, the ancient Eskimo principle, using an inner fur layer with the hair (pile) facing the skin, and an outer layer with the hair (pile) facing out. The Eskimo assembly was covered by a windbreaker made of dried walrus intestines.

Working in a cold environment may involve extremely heavy physical activity, alternated with rest periods. Since there is no single item of clothing capable both of protecting against cold at rest and of facilitating heat dissipation during heavy work, there is no single solution to this problem. A person sleeping outdoors at $-40\,^{\circ}C$ needs protection in the form of clothing and a sleeping bag equivalent to about 12 Clo units, i.e. 12 times that provided by the clothing which a person usually wears at room temperature. This is the insulating value of the original two-layer caribou clothing worn by the Eskimo (Åstrand and Rodahl, 1986). So far there is no clothing assembly available to modern man which may equal that of natural furs, although it should indeed be feasible to accomplish this with the artificial fiber material now available.

The importance of protecting the hands against heat loss in the cold is

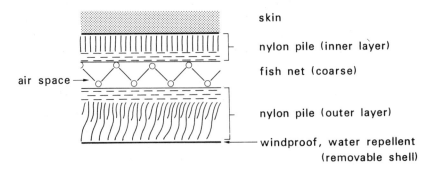

Figure 11.6. An example of a combination of fabric made of artificial fiber simulating the ancient Eskimo principle, using an inner fur layer with the hair (pile) facing the skin, and an outer layer with the hair (pile) facing out.

evident from the fact that although the surface of the hands only represents about 5 per cent of the total surface area of the body, about 10 per cent of the heat produced by a nude individual may be eliminated through the hands. In a clothed individual, up to 20 per cent of the heat produced may be eliminated through the hands (Day, 1949).

Roughly speaking, the insulation value of most materials is proportional to the amount of air that is trapped within the material itself. The reason is that air has an exceptionally high insulation capacity. Since the insulating value is primarily a function of the amount of trapped air in the clothing, the type of fiber used, for all practical purposes, is of less importance.

In a series of experiments, using similar garments made of natural wool pile, and of artificial fiber (nylon pile), it was shown that the artificial fiber was as effective in terms of providing thermal insulation and moisture permeability as the garments made of natural wool (Rodahl *et al.*, 1974; Figure 5.6 in Chapter 5). In addition, the garment made of artificial fiber was easier to clean, in that it could be washed in a regular clothes washing machine without shrinking. The main disadvantage with garments made of artificial fiber is that they are inflammable.

For practical purposes in the field, a physiological comparison of the insulating value of clothing may be obtained simply by measuring heat production (oxygen uptake), and heat loss as evidenced by changes in stored body heat (S), by applying the formula and procedures for obtaining rectal and mean skin temperatures described in Chapter 5. Evaporative weight loss may roughly be determined by weighing the subjects in the nude before and after the experiment. The accumulation of moisture in the experimental clothing may be assessed by weighing the garments before and after the experiment on a scale with an accuracy of ± 10 g.

The application of this approach to the evaluation of vapour permeability through different kinds of under garments may be illustrated by the following examples. For the purpose of comparing the moisture permeability of the so-called Lifa Super underwear made of thin threads of artificial fiber loosely woven, with other similar garments, such as a 'fish net' vest made of cotton threads, a double layer underwear made of artificial fiber (Duplo) and thin loosely woven cotton underwear (Calida), a series of experiments were performed in a climatic chamber at room temperature ($20\,^\circ\text{C} \pm 1\,^\circ\text{C}$), relative humidity about 30–40 per cent, in two healthy male subjects. The subjects walked for 60 minutes on a motor-driven treadmill at a speed of $100\ \text{m}\cdot\text{min}^{-1}$, inclination $3\,^\circ$. They then rested, sitting in the same room (climatic chamber) for 15 minutes followed by another 60 minutes walk on the treadmill. They wore, in addition to the experimental clothing, short cotton underpants, cotton socks, cotton training overalls and jogging shoes. The subjects walked together on the treadmill, one of them serving as a control, wearing Lifa Super underwear.

The sweating rate was determined by weighing the subjects nude before and after the experiment, on a Toledo scale, accuracy about 10 g. Since there was no fluid or food intake, nor any weight loss in the form of urine or stools, the change in body weight could be used as a rough estimate of sweat production. Evaporative sweat loss was determined as the difference between the weight loss of the subject, and the weight gain of all the clothing items worn. Each garment was accurately weighed on a precision scale before and after the experiment. The increase in weight was taken as an expression of the sweat accumulation in the garment. The heart rate, continuously recorded by a Vitalog data logger, was used as a control of the consistency of the rate of exercise on the treadmill. Body temperature, i.e. rectal temperature and skin temperature at the medial aspect of the thigh, were continuously recorded by the same Vitalog. On this basis the changes in body heat content (S) were calculated. Subjective ratings of the feeling of warm-cold and dry-moist were recorded before the start of the experiment, at rest and at the end of the experiment, using the Borg scale.

It was observed that more moisture was accumulated in the upper garment (the vests) than in the lower garments (the trousers). The moisture accumulation in the Lifa Super was small and fairly constant in the five repeated experiments. The moisture accumulation was less in the Lifa Super, the fish net and the cotton underwear, than in the Duplo heavier and denser underwear. This suggests that it is the loosely woven fabrics that allow moisture to escape, and not the kind of fiber itself which determines the moisture permeability of the garment. There was no consistent difference in body temperature or heat content which may be attributable to differences in the garments. There was very little difference between the experimental garments in the subjective ratings of comfort, cold or wetness.

From this simple test it was concluded that significant differences in moisture transfer were observed in the tested garments; Lifa Super, fish net and the Calida cotton underwear being superior in this respect to the Duplo tightly knit and heavier underwear. This finding supports the view that it is the way the fibers and threads are used and knitted, and not the fibers themselves, which primarily determine the moisture permeability. The spacing of the fibers seems to be the most important factor in this respect. This point was confirmed in a further experiment, the purpose of which was to compare the moisture uptake in different types of underwear, the rest of the clothing being the same in all experiments. Lifa underwear made of artificial fiber was compared with identical Lifa underwear made of wool. This was finally compared with ordinary woollen Stillongs underwear.

Three healthy young men were used as subjects. They wore one of the three experimental underwear, in addition to a nylon pile suit, an airline outerwear, a woollen cap, mittens, socks and shoes. They walked for one hour on the treadmill at a speed of $100 \, \text{m} \cdot \text{min}^{-1}$, at an inclination of $3°$ in a climatic chamber at a temperature of $-10°C$. Immediately following the exercise, the subjects rested for one hour, seated in the climatic chamber at $-10°C$. The observations included nude body weight before the experiment, after one hour's walk, and after one hour's rest. Moisture accumulation in the garments was determined by weighing each individual clothing item on a precision scale before the experiment, after one hour's walk, and again after one hour's rest.

A summary of the results is presented in Table 11.3. There was no statistically significant difference in the amount of sweat produced, regardless of the garment combinations used in the experiments. Nor was there any significant difference in the amount of the produced sweat which accumulated in the total clothing assembly (or the total amount of the produced sweat which evaporated), regardless of the garment combination used in the experiments. The fraction of the produced sweat which was picked up by the garments was roughly the same in the nylon pile suits (6–9 per cent), in the windbreaker (7–8 per cent), and in the cap, mittens, socks and shoes combined (6–7 per cent). In contrast, there was a marked difference between the Stillongs underwear and the Lifa underwear in terms of moisture accumulation: 6 per cent (or 38 g) in the Stillongs, as against 0.5–1 per cent (2–5 g) in the Lifa underwear. This difference was statistically significant ($p < 0.01$). There was no statistically significant difference, however,

Table 11.3. *Sweat lost and accumulated in the clothing during a 60 min walk on a treadmill followed by a 60 min rest in a climatic chamber at $-10°C$, expressed in g and per cent, using one of three different kinds of underwear (Stillongs, Lifa made of artificial fiber, and Lifa made of wool).*

	Stillongs wool	Lifa made of:	
		artificial fiber	wool
Sweat loss	606	566	550
Weight gain of:			
underwear	38 (6%)	2 (0.5%)	5 (1%)
Nylon pile suit	52 (9%)	41 (7%)	31 (6%)
Airline windbreaker	45 (7%)	46 (8%)	44 (8%)
cap, mittens, socks and shoes	36 (6%)	42 (7%)	32 (6%)
Total weight gain	171 (28%)	131 (23%)	112 (21%)
Sweat lost − sweat accumulated (i.e. amount evaporated):	435 (72%)	435 (77%)	438 (80%)

between Lifa made of artificial fiber and Lifa made of wool in terms of moisture accumulation.

Most of the moisture in the underwear was accumulated in the undervest, less in the underpants. Of the moisture accumulated in the rest of the clothing, about an equal amount was accumulated in the upper and lower garments (the jacket and the trousers). In the case of the other items, most moisture was accumulated in the shoes, followed by the socks, and then the mittens. The least amount was accumulated in the cap.

The obvious conclusions to be drawn from this rather simple experiment would be that Lifa underwear permits a greater transfer of moisture (in this case due to sweating) than does the Stillongs underwear. The fact that there is no significant difference between Lifa made of artificial fiber and Lifa made of wool suggests that it is not the type of fiber (artificial or wool) which determines the moisture permeability, but rather the manner in which the fiber (thread) is used in the garment. Of the underwear tested, the Lifa underwear maintains a drier microclimate next to the skin.

Equally important, however, is the ability of the outer garments to rid themselves of the evaporated sweat which has penetrated the underwear. In the present experiment some 70–80 per cent of the moisture was dissipated, i.e. some 20–30 per cent of the moisture produced remained in the clothing.

When the body, as a whole, is in a state of positive thermal balance, as in the case of prolonged physical activity, the peripheral parts, such as the hands, are also warm, because the body has surplus heat to dissipate. This is accomplished by relaxing the blood vessels in the peripheral parts of the body, including the hands, which thus become heated by the warm blood. Since manual dexterity is affected by the skin temperature of the hands, it is an advantage to remain sufficiently active so as to have surplus heat to get rid of through the hands. A typical feature of the primitive Eskimo hunter is that his physical work capacity (maximal oxygen uptake) is high enough for him to maintain a moderately high level of activity for prolonged periods (Vokac and Rodahl, 1976). Consequently, he is also able to keep his hands warm.

Cross-country skiing

Obviously, skis were originally made for the purpose of moving one's body and belongings across snow-covered terrain, especially across loose, deep snow, where walking on foot was extremely difficult or almost impossible. This is in contrast to moving through deep loose snow in wooded areas with dense undergrowth, where the use of snowshoes is preferable. Another advantage is that one may move much faster downhill on skis than on foot, and the work load may be somewhat lighter when sliding downhill (for references see Åstrand and Rodahl, 1986, p. 661). Brotherhood (1985) has presented an extensive review of the energy cost of travelling across snow, pointing out that travel at normal walking speed over any snow surface, regardless of mode, may be twice as energy-demanding as walking on a road. More muscle groups are engaged in skiing than in walking. Consequently, one would expect the overall energy expenditure involved in transporting the body on skis from one place to another to be as high as, or higher, than the energy expenditure when moving the body the same distance on foot.

Rønningen (1976) compared the energy cost of walking on foot on a hard, snow-covered level road, with skiing on a level trail alongside the road, using six young soldiers, dressed in military Arctic uniforms and carrying 22.5 kg on their backs. In both cases they moved at a speed of about 5 km per hour. He found identical values for oxygen uptake, i.e. about 1.5 liters O_2 per min. The subjects stated, however, that skiing felt easier than walking, probably because the same work load was being handled by a larger muscle mass in the case of the skiing, since the arm muscles were being used to aid locomotion through the ski-poles.

Skiing uphill at an incline of 7 to 8 per cent on a cross-country trail at a speed of about 4 km per hour required an oxygen uptake of about 2.3 liters per min on the average. One of the soldiers was studied during a winter manoeuvre, covering 120 km in five days across the mountains in Norway. Moving on skis with a load of 22.5 kg on the back, at a speed slightly less than 4 km per hour, required an average oxygen uptake of 1.8 liters per min, corresponding to roughly 40 per cent of the individual's maximal oxygen uptake. When two men pulled a 72 kg sled while on skis, the oxygen uptake increased to 2.3 liters per min on the average, corresponding to more than 50 per cent of the individual's maximal oxygen uptake. Christensen and Högberg (1950) compared the energy cost of carrying an extra load of 30 kg in a rucksack, and the transportation of the same load on a sled. The oxygen uptake in the rucksack experiment was only 70 to 80 per cent of the oxygen uptake required to transport the same load on a sled at a speed of 4 km an hour.

In view of the increased energy cost of moving across snow-covered terrain, one would expect that living and working in the Arctic (or the Antarctic) would require greater caloric expenditure, and hence increased caloric intake than is the case in more temperate zones. Actually, this may be the case, the main reason being the hobbling effect of the cold weather clothing (Consolazio, 1963), in addition to the extra cost of moving across snow-covered, difficult terrain. In either case, there is little or no change in basal metabolism, and indoor activities in the Arctic or Antarctic require the same energy expenditure as in temperate or subtropical areas (for references, see Brotherhood, 1973).

Chapter 12
The physiology of fishing

According to Eaton and Konner (1985), widespread use of aquatic food appears to be a fairly recent phenomenon, since fish bones and shells are infrequently found in archeological material older than 20,000 years. It may therefore be safe to conclude that fishing is a fairly recent occupation, compared to hunting. Nevertheless, in modern times, both coastal and deep-sea fishing have played a major role as a basis for habitation and subsistence in maritime countries, such as Norway. Yet, this occupation has not been subject to as much occupational health interest as have land-based industrial occupations.

Of the available publications dealing with different aspects of fishing, the majority deals with studies pertaining to health (such as Newhouse, 1966; Kersten, 1967; Spitzer *et al.*, 1975; Cadenhead, 1976; Fugelli, 1979a, b: Gaddie *et al.*, 1980; Cross, 1985; Luksza, 1985; Wattkins, 1986), and safety, including health hazards, accidents and risks of different kinds (Ellis, 1970; Dalgaard *et al.*, 1970; Schilling, 1971; Dalgaard *et al.*, 1972; Zugaj and Chmielewski, 1974; Collacott, 1977; Vanggaard and Nielsen, 1977). Helgason *et al.* (1977) have described psychosocial aspects of the life of deep sea fishermen and their families, and compared them with the problems of industrial factory workers and their families. Although it is generally agreed that fishing is an arduous occupation, only a limited number of publications deal with specific physiological aspects of the fishing occupation, including work stress, work load and ergonomic aspects.

Bondarev *et al.* (1962) investigated the energy expenditure of fish processing personnel on board trawlers in the Barents Sea and the North Atlantic. Dobronrarova (1962) made a similar study among fishermen in Arctic waters. They found that the energy expenditure was high and that the load on the circulatory system was highest during the handling of the catch. They also observed that the energy expenditure increased noticeably with increasing wind and heavy seas. Sandrackaja (1966) reached similar conclusions and observed a reduction in the finger strength in fishermen who were engaged in the cleaning and preparation of the catch, which was attributed to local cooling of the hands. Nilsson (1970) studied the energy expenditure of Norwegian fishermen engaged in Danish seine fishing off the Western coast of Norway. On the basis of oxygen uptake and heart rate he estimated the work load to be between 30 and 45 per cent of the fisherman's aerobic power. His data indicated an increase in the work load in rough weather. His findings suggest that persons with low work capacity, especially older individuals, and persons with cardiovascular disease, may, under certain circumstances be working under considerable strain.

Information of this kind is of considerable practical and clinical importance, particularly in connection with evaluating a person's fitness for work in the different branches of the fishing industry, in assessing the need for early retirement, and in assessing disability in the case of fishermen with disabling diseases, especially those of the respiratory, cardiovascular and locomotor system. This was the reason for the Norwegian Institute of Work Physiology conducting a series of studies of work stress in the Norwegian fishing industry, involving both coastal and deep-sea fishing. Initially, these studies were carried out in collaboration with I. Åstrand et al. at the Swedish Institute of Occupational Health (I. Åstrand et al., 1973; Rodahl et al., 1974b; Rodahl and Vokac, 1977a, b).

It is understandable that comparatively few detailed studies have been made of fishermen working aboard small boats because of the considerable practical difficulties involved in the collection of accurate physiological data under field conditions during fishing operations in rough weather. In view of these practical difficulties, it was considered preferable to limit the observations to a smaller number of subjects rather than to examine a larger number more superficially.

Coastal fishing

A series of work physiological studies of 24 fishermen engaged in common types of coastal fishing at Lofoten (68° northern latitude) during a total of 35 working days was carried out during the fishing season in 1971 and 1972 (I. Åstrand et al., 1973b; Rodahl et al., 1974b). The first study was carried out during February and March 1971, under rather unfavourable weather conditions, with temperatures below freezing (-3 to $-9\,^{\circ}$C), wind velocities 5–20 meters per second, and rough to very rough sea. This study was supplemented by some observations on hand-line fishermen at Vaerøy, a small island west of the main Lofoten archipelago, in June 1971, under very favourable weather conditions (no wind, calm sea, and temperatures around $10\,^{\circ}$C). The second study in 1972 was conducted at Vaerøy during the months of March and April under slightly better weather conditions than during the first study, with temperatures between -2 and $+2\,^{\circ}$C, and wind velocities 3–5 meters per second. The fishing operations studied included the hand-line, long-line, net, and Danish seine. The *hand-line* is a nylon cord with about a dozen hooks supplied with permanent artificial bait. A weight is attached to the end of the line. It is lowered to a couple of fathoms from the bottom. The depth varies from about 40 to 70 fathoms. The fishing usually takes place from small boats, some 20 to 40 feet long, as a rule manned by one man, occasionally by two or more. The line is lowered and raised with the aid of a hand-operated wheel. During the summer study an automatic device was used, replacing the manual operation by an electrically operated wheel that lowered and raised the line.

The *long-line* consists of a thick cord more than 300 fathoms long, to which the hooks are attached with the aid of short thin nylon strings, about a fathom apart, so that each long-line has about 350 hooks which are baited with shrimp, mackerel, etc. Several such long-lines may be tied together, one after the other, and lowered horizontally into the water along the bottom of the sea. For this type of fishing the boats used are about 35–45 feet long and manned by a crew of 2–3 men.

The *nets* which are made of nylon, are usually tied together, one after the other in a straight line, to form a long string of nets up to several hundred fathoms in length. They are placed near the bottom of the sea, usually at a depth of 40 to 60 fathoms. The 50–70 foot long boats used for this purpose are usually manned by a crew of 6–7 men.

The *Danish seine* consists of a very long net. It is placed in the water in a wide circle surrounding a school of fish. The seine is pulled together by the power block until the catch is finally held in a portion of the net like a bag, which is then hoisted on board by a winch. The boats used for this type of fishing are usually about 50–80 feet long and manned by 3–4 men.

All subjects studied were full-time, professional fishermen, who had been active in this occupation since they were about 15 years old. In the first study, 14 fishermen were studied for the entire working day on board the boat. In the second study, ten other fishermen were studied while working on board their boats out at sea, four of these during one whole day, two of them for two days, and five of them for three consecutive days. In addition, eight catch-handlers working ashore were studied during the second study (1972), two of them for three consecutive days. The average age of the 24 subjects was 48 years (18–69). At the time of the study the mean age of all fishermen in Norway was 46 years. For the eight catch-handlers the mean age was 57 (25–82). A tendency towards overweight with increasing age was noted in both the fishermen and the catch-handlers. Eight out of 20 subjects over 50 years of age were 15–45 per cent heavier than the predicted values from the height (Jelliffe, 1966), while all 12 subjects under 50 years of age were within ± 15 per cent of the weight predicted from the height. Their state of physical fitness ranged from fair to good, as judged by their maximal oxygen uptake (roughly 2–3.5 l $O_2 \cdot min^{-1}$).

A mechanically-braked cycle ergometer was used to determine the heart rate response to known work loads, engaging large muscle groups, starting at 300 $kpm \cdot min^{-1}$, and increasing stepwise every 6 min until a heart rate of 160 $beats \cdot min^{-1}$ was attained. After a few minute's rest the maximal oxygen uptake was determined by the Douglas bag method. Peak blood lactate concentration was determined 2 min after the cessation of the maximal effort.

For the purpose of translating recorded heart rates in the field into predicted oxygen uptake (or work rate), the individual relationship between heart rate at different submaximal work loads and the corresponding oxygen uptake was plotted graphically (see Chapter 4). On the basis of the continuously recorded heart rate in the field, the approximate oxygen uptake during work could then be predicted and expressed in per cent of the individual's maximal aerobic power. During the field studies, a member of the research team went out with the boat to keep detailed records of the sequence and duration of all activities in which the subject, under study, was engaged.

The work load imposed by the different fishing operations was assessed both by measuring the oxygen uptake during the various fishing operations and by recording the heart rate continuously. The oxygen uptake was measured by collecting expired air in Douglas bags during typical work situations, while the heart rate, at the same time, was continuously recorded. The volume of the expired air was measured on board the fishing vessel by a dry spirometer, and samples of the gas were analyzed in the laboratory ashore. The subject's heart rate was continuously recorded during the entire work-day with the aid of a portable, battery-driven tape recorder. On this basis, the work load to which the subject was exposed was assessed.

For the assessment of the total work stress, to which the subject was exposed, urinary catecholamine excretion was determined in urine samples collected from 0600 to 1600 hours, compared with night urine samples collected from 2200 to 0600 hours. As an additional check on the urinary sample collection, urinary creatinine excretion was determined.

The work loads of a series of typical fishing operations, in terms of actually measured oxygen uptake in the field, are presented in Figure 12.1. From this figure alone, it is evident that a variety of factors may affect the oxygen uptake required in one and the same task, apart from the marked difference in oxygen uptake between the different tasks. The lowest oxygen uptake was observed when fishing with the hand-line in a sitting position (subject no. 1). The simple task of maintaining balance when fishing with the hand-line in a standing position increased the oxygen uptake considerably (subjects nos. 2, 3 and 4). In fact, the extra energy cost of keeping the balance when standing on deck in heavy seas is quite considerable, and may, in the long run, largely affect energy balance.

The weight of the catch is another factor. Figure 12.1 shows a noticeable increase in the oxygen uptake when handling the hand-line with many fishes on it (subjects nos. 8 and 9). The increase in the oxygen uptake due to the weight of the catch is also well illustrated in subject no. 11, pulling in the net with many fishes, compared to when pulling in an empty net.

Steering the boat requires a low oxygen uptake (subject no. 10), as does standing ashore putting baits on the hooks on the long-line (subjects nos. 15, 16). The highest oxygen uptake was repeatedly observed when pulling in the Danish seine by power block, when the oxygen uptake (2–2.7 l O_2 per min) reached 60–80 per cent of the maximal aerobic power of the fisherman (subject no. 13). This activity, which occurred every hour or two, lasted for a few minutes, altogether

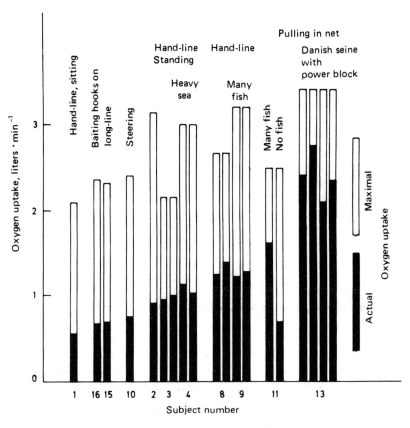

Figure 12.1. Measured oxygen uptake in typical fishing operations (from I. Åstrand et al., 1973b).

for some 20 minutes a day. A physical work load of this magnitude and duration may be sufficient to cause a training effect in the subject, tending to increase his maximal aerobic power. It is of interest to note that the highest aerobic power was actually observed in one of these Danish seine fishermen, 48 years of age.

Contrary to the typical, short periods of high level exertions in Danish seine fishing, net and long-line fishing usually require a somewhat lower, but more prolonged energy expenditure. They may, therefore, in the long run, be even more exhausting. This is also in agreement with the subjective feeling of fatigue of the fishermen themselves.

On the whole, hand-line fishing may, under certain circumstances, impose the lowest load on the fisherman. Furthermore, the hand-line fisherman may operate on his own, opposed to long-line, net and Danish seine fishing, where each man is obliged to keep up with the rest of the team. In addition, the hand-line fisherman is, to a greater extent, free to choose whether or not he should stay at home if the weather is too rough. The long-line and net fishermen are under greater pressure to get out even under rather extreme weather conditions, in order to save the catch and the fishing gear, which otherwise may drift away or become destroyed. For these reasons, hand-line fishing is probably the most suitable type of fishing operation for older fishermen, and for those who have any medical or physical handicaps.

The energy cost of the different activities on board the different types of fishing boats in the course of a working day, estimated on the basis of the continuously recorded heart rate, is listed in Table 12.1. The average estimated energy expenditure of all activities on board during the whole day amounted to the equivalent of an oxygen uptake of 0.9–1.1 l per minute. This corresponded to 34–39 per cent of the fishermen's maximal aerobic power, with occasional peaks of up to 80 per cent. The heart rate exceeded 50 per cent of the fishermen's heart rate reserve ('heart rate reserve' being the difference between maximal and resting heart rate) for 9–23 per cent of the observation period. The most strenuous activities were pulling in the seine with a power block (oxygen uptake up to $2.7 \, \mathrm{l \cdot min^{-1}}$) and unloading the catch, taxing the subjects by more than 50 per cent of their maximal aerobic power for two-thirds of the duration of these activities. The intermittent character of practically all activities on board was a typical finding.

A striking feature of the fisherman's profession is variability, depending on weather, amount of catch and other unpredictable circumstances. This also includes the length of the working day out at sea, which in this study varied from 2.5 to 15 hours. Another feature is the flexibility in work performed. The skipper, for instance, not only steers the boat, but may also give a hand pulling in the seine. Job rotation is a must. Any member of the crew has to be able to steer the boat, handle the

Table 12.1. *Work load, assessed on the basis of heart rate, in different types of coastal fishing. Means and range in parenthesis (from Rodahl et al., 1974b).*

Type of fishing	Type of activity	No. of subjects	Estimated oxygen uptake liter/min STPD	Per cent of maximal aerobic power	Heart rate >50% level Duration, min
Hand-line	Operating line	7	0.9 (0.5–1.6)	36 (20–60)	22 (0– 82)
	Whole observation period	7	0.9 (0.5–1.6)	37 (20–60)	44 (1–136)
Long-line	Pulling in line	5	0.9 (0.6–1.4)	35 (25–50)	18 (0–106)
	Putting out line	5	0.8 (0.7–1.2)	31 (30–40)	1 (0– 2)
	Whole observation period	8	0.9 (0.5–1.4)	34 (25–45)	30 (3–150)
Net	Pulling in net	5	1.2 (1.0–1.4)	40 (30–55)	27 (1– 87)
	Putting out net	2	1.2 (1.0–1.5)	47 (35–60)	12 (1– 23)
	Whole observation period	6	1.0 (0.8–1.2)	35 (25–50)	45 (22– 85)
Danish seine	Pulling in seine	2	1.7 (1.6–2.0)	56 (55–60)	97 (51–142)
	Putting out seine	2	1.4 (1.3–1.5)	45 (45–45)	8 (0– 15)
	Whole observation period	3	1.1 (1.0–1.4)	39 (35–45)	109 (59–152)
Common for all types	Steering the boat	13	0.9 (0.4–1.4)	37 (15–55)	6 (0– 54)
	Unloading catch	4	1.4 (0.8–1.6)	43 (25–65)	8 (0– 34)
	Bleeding and cleaning catch	10	1.1 (0.6–1.8)	40 (30–60)	6 (0– 20)
	Other activities	21	1.0 (0.5–1.8)	39 (25 70)	15 (0– 66)
	Resting	18	0.7 (0.3–1.1)	25 (10–45)	0

fishing tackle, bleed and clean the catch. Yet in spite of the varying incidence and duration of the different activities, the average work load during the entire work day was remarkably alike in all four types of coastal fishing studied, i.e. about 1 liter oxygen per minute (330 kpm·min^{-1} on a cycle ergometer), which corresponded to 34–39 per cent of the maximal aerobic power of the subjects. However, fishing with a hand-line, as already mentioned, required less energy expenditure than handling the net and, especially, the Danish seine. Unloading, as well as bleeding and cleaning the catch, belongs also to activities involving rather high average energy expenditure.

On the other hand, many of the activities on board are of an intermittent character, i.e. high levels of energy expenditure are interspersed with recovery periods in the form of breaks or less demanding activities. In addition, resting periods during sailing to and from the fishing grounds, and when waiting between the actual fishing operations, amounted to approximately one-third of the work-day.

In Chapter 4 it was suggested that the heart rate reserve (HRR) concept (maximal heart rate minus resting heart rate), may be a useful index of circulatory strain. As a reasonable upper limit, in the case of prolonged work, a work load corresponding to 50 per cent of the heart rate reserve may be acceptable, based on practical experience. (The 50 per cent level is defined as the resting heart rate + 50 per cent of the difference between maximal and resting heart rate.) Both pulling in the seine and unloading the catch caused the heart rate to exceed this upper limit for two-thirds of the duration of these activities.

It is evident from the present study that even a task such as steering the boat required, on the whole, a comparatively high oxygen uptake of about $0.9 \, l \cdot min^{-1}$ (37 per cent of the maximal aerobic power), which is considerably more than the oxygen uptake when standing still on land. This is probably due to the counter-balancing of the motion of the boat which requires considerable muscular effort. This may explain the comparatively high energy expenditure values recorded in our fishermen during rest.

Two hand-line fishermen were studied both during the winter when the weather was rough, and during the following summer when the weather was calm. The average heart rate and oxygen uptake in both of them were markedly lower in the summer than in the winter (0.8 as against $1.3 \, l \cdot min^{-1}$).

The study of the eight men working as catch handlers ashore during a total of 12 working days, showed that the handling of the catch entailed a mean oxygen uptake of $0.98 \, l \cdot min^{-1}$, corresponding to about 34 per cent of their maximal aerobic power. The peak values were con-

Table 12.2. *Work load, assessed on the basis of heart rate, in catch-handlers on land. Means and range in parentheses (from Rodahl et al., 1974b).*

Type of activity	No. of subjects	Estimated oxygen uptake, liter/min STPD	Per cent of maximal aerobic power	Heart rate >50% level Duration, min
Hanging fish	2	1.1 (1.0–1.2)	40 (35–45)	1 (0– 2)
Cutting fish and removing bones	3	0.9 (0.7–1.0)	33 (25–50)	1 (0– 6)
Salting	2	0.9 (0.8–1.0)	37 (35–40)	3 (0– 6)
Preparation of roe with sugar and salt	1	1.4	55	64
Manual transport	3	1.2 (1.0–1.4)	39 (35–55)	14 (2–40)
Motorized transport	2	0.8 (0.7–1.3)	23 (20–40)	2 (0– 4)
Other activities	8	0.9 (0.7–1.7)	35 (25–70)	10 (0–52)
Resting	8	0.6 (0.4–0.8)	23 (10–40)	0

siderably lower and of shorter duration than in the fishermen (Table 12.2.).

The average urinary excretion of epinephrine and norepinephrine during the working day was high in both the fishermen and the catch-handlers (72 ng·min^{-1} norepinephrine and 24 ng·min^{-1} epine-phrine in the fishermen, and 59 ng·min^{-1} norepinephrine and 15 ng·min^{-1} epinephrine in the catch-handlers). In the case of the fishermen, a nearly ten-fold increase in the epinephrine excretion, and about a four-fold increase in the norepinephrine excretion were observed during the work day as compared with the excretion during the night. The difference was statistically highly significant ($p < 0.001$). The epinephrine excretion in the fishermen was significantly higher than in the handlers ($p < 0.01$).

The highly significant increase both in epinephrine and norepine-phrine excretion in the fishermen is probably due to several factors: the excretion is generally higher during the day than during the night, due to circadian rhythmic variations. Body posture also affects catechol-amine excretion; it is higher when standing than when lying (Sundin, 1956). Cold exposure increased catecholamine excretion (Lennquist, 1972). Catecholamine excretion is elevated during physical stress (Frankenheuser, 1968). Finally, emotional factors also contribute to the increased excretion. It should be pointed out, however, that the catecholamine levels reported here, especially of epinephrine, are at least twice as high as the values which have been found in investigations with emotional stimuli alone (Levi, 1967).

Trawler fishing

Work stress associated with modern trawler fishing was investigated on board two medium-sized (298 GRT) Norwegian stern trawlers (Rodahl and Vokac, 1977a). One of the trawlers was operating off the coast of Labrador, catching Greenland halibut; the other was operating in the Barents Sea, north of the coast of Finnmark (Norway) catching cod, haddock, etc. The circulatory strain and the work load were assessed by computerized analysis of the continuously recorded heart rate in six subjects working in regular six hour shifts. In both trawlers it appeared that roughly one third of the time was spent working, one third resting and one third sleeping. The duration of actual work in four deck hands and a trawler foreman amounted to 33–41 per cent of the 24-hour schedule, and taxed, on the average, 30–38 per cent of their heart rate reserve (equivalent to an oxygen uptake of 0.9 – 1.1 liter per minute) with peaks up to 80 per cent. The mean work load of a skipper, whose work was sedentary, corresponded to 20 per cent of his heart rate reserve. The regular pattern of activity, as well as the average energy expenditure, were quite similar in the two trawlers examined. There was also a high degree of reproducibility in the circulatory strain in the same subject in different on-duty shifts. The average energy expenditure for a 24-hour period as well as the urinary catecholamine excretion rates were lower than in coastal fishermen. It is evident that apart from the long absence from home, the work on board a modern trawler can, in many respects, be quite similar to many ordinary work places ashore.

One of the main differences between coastal fishing and trawler fishing is that the work on board the modern trawler is mechanized to a large extent, although a considerable amount of manual work is still necessary. Furthermore, while the coastal fishermen, as a rule, return home for the night, the fresh fish trawlers are out at sea for about 12 days at a time. Some of them, like one of the trawlers included in the present study, may not return to the home port for months, creating the kind of problems described by Helgason *et al.* (1977). This prolonged absence from home may, in fact, represent one of the greatest disadvantages of modern trawler fishing. In an attempt to alleviate this hardship, a system was introduced on board the Barents Sea trawler by which, on each trip, one of the crew remained at home on leave.

Bank fishing

A study was performed on board two long-line fishing ships: a modern Faeroe Island trawler with a crew of 15 fishing Greenland halibut off the coast of Labrador, and a Norwegian vessel with a crew of 11 fishing dogfish off the Orkney Islands (Rodahl and Vokac, 1977b). Three men from the crew of the Faeroe Island trawler and two men from the Norwegian trawler served as subjects for this study.

The average work load of the three deck-hands on board the Faeroe Island vessel, as judged by their mean circulatory strain, was moderate (25–33 per cent of the heart rate reserve). However, the average 9–18 min duration of a heart rate exceeding 50 per cent of the heart rate reserve, together with the peak heart rates of up to 165 beats/min, indicates significant periods of intense physical strain, especially when unhooking the fish. On the whole, these work loads are similar to, though slightly lower than, those observed in deck-hands on board trawlers and among long-line fishermen engaged in coastal fishing.

In general, these observations were confirmed by the findings made on board the Norwegian vessel. Here, it was shown that storing the catch on ice, a rather strenuous operation, caused heart rates of around 150 beats/min continuing for over an hour. However, the work has a rhythmical, intermittent character with 3–4 min periods of intense effort followed by about the same duration of less intense activity.

By far the most strenuous operation is the unhooking of the fish as the line is pulled in by winch over the side of the ship. It is generally

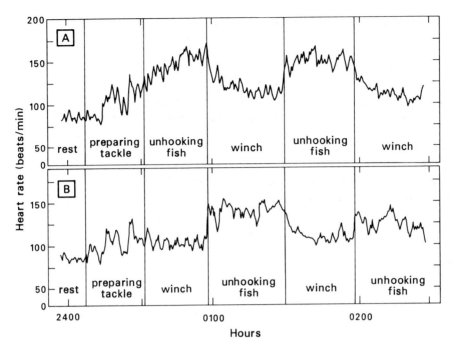

Figure 12.2. Heart rates in two deck hands on board a Norwegian long-line fishing boat, fishing dogfish off the Orkney Islands, changing places at regular intervals between unhooking the fish and operating the winch.

considered that at times when the fishing is good and there is a dogfish on almost every hook, this particular operation cannot be endured for more than about 25 min at a stretch. Consequently, a routine has been established whereby the deck-hand unhooking the fish and the one operating the winch change places at regular intervals. The mean heart rates when unhooking fish were 154 (range 115–190) in one subject and 134 (105–155) in the other subject. The corresponding mean heart rates when operating the winch were 130 (105–160) and 113 (95–140) respectively. (Figure 12.2). The urinary catecholamine excretion rates in the bank fishermen studied were of the same order of magnitude as those observed in the trawler fishermen. Periods of extremely high work stress were observed on board the Norwegian trawler, when the fishing was exceptionally good, and the crew had to work several days running without sleep because of the reduced number of the crew.

Capelin fishing

Three crew members (skipper, deck-hand, and cook) of a Norwegian purse-seine fishing vessel fishing capelin off the coast of Finnmark, were studied. The field study included a 24-hour continuous observation period during a trip lasting altogether 72 hours. Due to the remarkable efficiency of the methods developed for capelin fishing, the vessel may be filled to capacity in a matter of a few hectic hours. The rest of the time is spent waiting, searching or sailing to and from the port. Consequently, the circulatory strain observed in the skipper was slight, with a mean heart rate of about 90 beats per min. However, during the critical periods of the actual fishing operations, his heart rate reached peak values of 155 beats per min, and for a period of 14 min exceeded 50 per cent of his heart rate reserve.

In the case of the cook, his mean heart rate was 90 beats per min (ranging from 65 to 130) corresponding to an average work load which taxed about 15 per cent of his heart rate reserve. On only a few occasions did his heart rate exceed 50 per cent of the heart rate reserve.

The 23-year old deck-hand who served as the helper to the purse-seine foreman, was subject to exceptionally high work loads during the actual fishing operations, with peak loads exceeding 50 per cent of his heart rate reserve lasting for a total of 100 min out of the 24-hour observation period. Otherwise he spent about half of the time resting, with a heart rate ranging from 50 to 90 beats per min. The remaining time he was occupied with miscellaneous work on board and standing at the wheel, taxing roughly 15 per cent of his heart rate reserve.

Energy expenditure in modern fishermen compared to primitive Eskimo seal hunters

From these studies it is evident that fishing may entail hard physical work. Yet, if properly organized, with job rotation and an adequate number of crew members to share the work, it may, in some respects, be similar to many ordinary occupations ashore.

Much of the improved efficiency of the fishing operations is due to the introduction of mechanization and automation. This development has led to an increased tempo, imposing a greater stress on the crew at times. This is especially the case in the older fishermen who, on account of their declining physical work capacity, may find it difficult to keep up with their younger colleagues when the fishing operations are carried out by a team, as in the case of long-line, net or Danish seine fishing. Yet, the official Norwegian statistics show that the percentage of the older fishermen is increasing. At present, almost half of them are over 50 years of age.

In view of the above mentioned trends in the development of the modern fishing industry, it might be of some interest to compare the work stress of the present-day fisherman with that of some of the remaining primitive Eskimo seal hunters in Greenland. During the studies of the long-line and trawler fishing off the coast of Labrador the opportunity presented itself to visit an Eskimo settlement on the north-west coast of Greenland, and to make a study similar to that carried out on the fishermen on a group of Eskimo hunters (Vokac and Rodahl, 1976).

The investigation was carried out in September 1973 in Kraulshavn (population about 100), latitude 74° north, on the NW coast of Greenland. Eight active Eskimo hunters (about one-third of all the hunters in the village) volunteered for the study. Four of the subjects had participated in a similar investigation in 1969 (Lammert, 1972). A laboratory was set up in the Eskimo village. The subjects were patiently taught how to use a Monark cycle ergometer, and given ample time for practice, since bicycling is an exercise entirely unfamiliar to the Eskimo. The subjects then exercised both at submaximal and maximal work loads, in the first case for 7 min each at 50 and 100 watt, pedalling at 50 revolutions per min on the mechanically-braked Monark cycle ergometer. Heart rates were recorded on a conventional electrocardiograph. Pulmonary ventilation and oxygen uptake were measured by the Douglas bag method, using a dry spirometer and a micro-Scholander analyzer. Peak blood lactate concentration was determined from capillary blood collected from the pre-warmed finger tip, 5–6 minutes after the cessation of maximal exercise.

The field studies included assessment of circulatory strain during a 24-hour period, including a usual day's work of hunting seals from a combination of motor boats and kayaks. This was done by recording the subject's heart rate continuously with the aid of portable, battery-operated Hellige tape recorders. Detailed records were kept during the hunting trips of the sequence and duration of the activities of the subjects. Quantitative evaluation of the heart rate was carried out with the aid of a digital computer as described in detail by Rodahl *et al.*, (1974). In addition, urinary catecholamine output (epinephrine and norepinephrine) was assayed (Vaage, 1974) from urine samples collected between 0800 and 2000, and 2000 and 0800 hours.

The maximal aerobic power in terms of maximal oxygen uptake measured on the cycle ergometer was exceptionally high in 6 of the Eskimos, and average in two of them (Table 12.3) compared to the physical fitness classification commonly used for Caucasian subjects (Åstrand and Rodahl, 1986). This is in contrast to the rather low values reported earlier for Eskimos of the same community (Lammert, 1972). The four subjects who participated in both studies showed a 52 per cent higher maximal oxygen uptake in the present study compared to the study referred to in 1969 (for details see Vokac and Rodahl, 1976).

The very high level of physical fitness in a majority of the Eskimo subjects of the present study is similar to the Canadian Eskimos examined by Rode and Shephard (1971). The explanation for this may be found in the intermittent high work loads encountered during hunting and other daily activities, as is evident from Figure 12.3. Although the periods of intense physical activity may be of short duration, they may tax up to 80 per cent of the heart rate reserve and last long enough to produce a significant training effect. In the present study the average duration of heart rate over 50 per cent of the heart rate reserve amounted to half an hour daily, and is probably enough to produce a training effect. Since short bouts of intense activity are more effective in developing physical fitness than long periods of moderate effort (Åstrand and Rodahl, 1986), the intermittent high circulatory strain would explain the level of physical fitness observed during the summer hunting season. It may be assumed that the long, arduous sled journeys during the winter would help to maintain or even augment a superior physical fitness.

The field observations indicate that during the last half hour on shore, the subjects were occupied by rather heavy physical work, consisting of loading their motorboats with fuel, provisions and other equipment, which had to be transported to the boats from the houses situated on the hills above the harbour. This effort is clearly reflected in the intermittently high heart rates, with peaks of up to 150 beats/min. Similar high heart rates with even higher peaks were recorded during the first half hour on shore upon return from the hunting trips (subject 2, Figure 12.3).

The low mean heart rate in the long periods of cruising along the shore or waiting for the seals reflects the general inactivity on board the motorboats. Short periods of activity and moderate physical exertion (peak heart rates 120–140 beats/min) were noted when the subjects

Table 12.3. *Physical characteristics and maximal oxygen uptake, heart rate and lactic acid of the Eskimo subjects tested in the laboratory (from Vokac and Rodahl, 1976).*

Subject no.	Age, years	Ht, cm	Wt, kg	$\dot{V}O_2$ ml/kg·min	RQ,	Lactic acid, mM/l	Heart rate, beats/min
1	30	173	69	61	1.01	11.6	183
2	30	160	61	57	1.08	9.8	197
3	34	164	70	62	1.09	10.4	191
4	34	163	59	60	1.12	11.6	165
5	35	162	62	45	1.10	5.0	165
6	37	170	69	41	1.09	7.6	186
7	48	160	55	52	1.03	8.6	157
8	51	167	57	50	0.94	6.6	170
Mean	37.4	164.9	62.8	53.4	1.06	8.9	176.8
SE	2.8	1.7	2.1	2.8	0.02	0.8	5.1

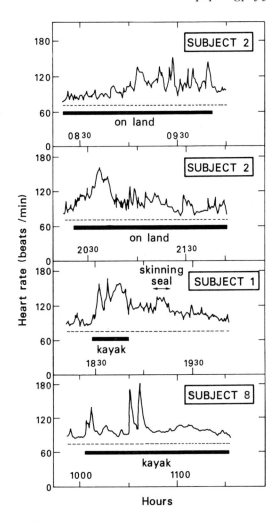

Figure 12.3. *Continuous heart rate recordings in three Eskimo subjects during typical periods of activity, including seal hunting. The resting heart rates are indicated by dashed lines (from Vokac and Rodahl, 1976).*

sighted a seal and shot at it, when they let down and hoisted up the kayaks, or when they hoisted up and skinned the seals (subject 1, Figure 12.3).

The highest peak heart rates were recorded when paddling in the kayaks in pursuit of a wounded seal (subject 1, Figure 12.3). Sometimes the subjects preferred to hunt from the kayaks in which they might stay for over two hours at a stretch. The low mean heart rates during these periods indicate that they were mostly sitting and waiting for a seal to appear, with only occasional slow paddling alternating with short bouts of extreme effort (subject 8, Figure 12.3).

The mean heart rate during the periods on shore, before and after the hunting trips, was not especially high (median 27 per cent HRR), although the peak heart rates, reaching 64–82 per cent HRR, and the duration of a heart rate of over 50 per cent HRR (12 min on average) confirm the impression of acute periods of circulatory strain before and after the hunting trips. When the hunters had finished unloading the motorboats and disposed of the seals and the equipment (30–40 min) they stayed awake until late at night (11 p.m.–1 a.m.). The heart rates during these periods were generally quite low (10–15 per cent HRR). The subjects then slept for 6–9 hours.

The mean urinary catecholamine elimination during the day was about 42 ng/min for norepinephrine and 17 ng/min for epinephrine. The corresponding night values were 26 and 8 ng/min. The difference between the day and night norepinephrine elimination rates was not significant ($p > 0.05$), while the epinephrine values differed at a significant level of $p < 0.025$. These values are considerably lower than those recorded in Norwegian coastal fishermen, and are comparable to the levels usually observed in industrial and other manual workers (Vokac and Rodahl, 1975). This implies a comparatively low degree of stress in Eskimo hunters, compared to those engaged in the modern fishing industry.

Chapter 13
Stress at sea

Introduction

The occupation of the sailor is unique, not only because of the long periods away from home, but also because of the isolated environment on board. Furthermore, it is an occupation based on a long history of well-established traditions, which is now subject to profound and rapid changes due to the introduction of new and advanced technology, and the resultant radical changes in the concepts of manning and operating ships, which have emerged in recent years. It is also an occupation presenting certain unique health and safety problems, including an exceptionally high suicide rate (Goethe et al., 1978a). The latter may perhaps partly be explained by the fact that a suicide attempt is more likely to be fatal out at sea, where anyone deciding to jump overboard is not very likely to be brought back alive. According to Ross (1977), referring to Otterland (1970), merchant seamen are seven times more likely to have fatal accidents than men of the same age ashore.

Dolmierski and de Walden (1972) examined the Polish records of fatalities at sea over the previous decade, including suicides and missing persons during voyages of ships from the merchant marine. The frequency of successful suicides amounted to over 20 per cent of the total number of cases. The most frequent causes of suicide were depression, psychotic or neurotic syndromes, as well as alcoholism. In the majority of the cases there had been a tendency to alcohol abuse, and many of the suicide victims had been under the influence of alcohol at the time of death. In their conclusion Dolmierski and de Walden (1972) recommended stricter control of alcohol intake by seamen. In this connection, it may be of interest to note that according to Elo (1979), neurosis and alcoholism have been the most noticeable mental disturbances found among seamen. Furthermore, a study of 220 drowning accidents in Singapore in 1973 to 1976, of which 44.5 per cent occurred at sea, showed that alcohol was found in the blood in over 20 per cent of all accidental drownings above 15 years of age (Ng et al., 1978).

The problems of the sailor have been subject to a considerable amount of interest from research workers of different kinds (Jonsen,

1976). The German Bernhard-Nocht Institute in Hamburg has com-
piled a collection of more than 6,000 publications dealing with different
aspects of nautical medicine and physiology. Members of the Japanese
Maritime Labour Research Institute in Tokyo have studied working
sailors at sea ever since the institute was established in 1966. These
studies have included the recording of the heart rate in captains during
manoeuvering in and out of ports. The results of these initial studies
were not readily available to most of us, since the reports generally
appeared in Japanese. In 1967, Ohashi and Sugihara showed how the
heart rate of a captain was affected by the mental tension caused by the
manoeuvering of his ship in and out of port. Ohashi and Hirota (1969)
monitored the heart rate of shipmasters and coastal pilots in nearly 100
cases since 1964 with the aid of telemetry, and were able to demonstrate, rather convincingly, how mental stress or difficult situations
which evoked mental tension, cause an elevation of the heart rate. As an
example, they showed the heart rate of the captain of a supertanker
during the passage of the narrow Kudako channel at Setonaikai in
Japan, during two critical situations which could have led to collisions
with small fishing boats, and where the captain intercepted and changed
the orders given by the pilot. They also presented typical examples of
elevated heart rates, up to 140 beats per minute, during manoeuvering
in and out of ports in heavy traffic.

Kilbom (1969) recorded the heart rate continuously in a Swedish
coastal pilot and showed that he reacted on the approach of another ship
by increased heart rate. Höyem-Johansen and Natvig (1978) studied the
work stress of four Norwegian coastal pilots under favourable weather
conditions. They did this by continuously recorded heart rate, paired by
the keeping of time-activity logs (Figure 13.1), and the analysis of

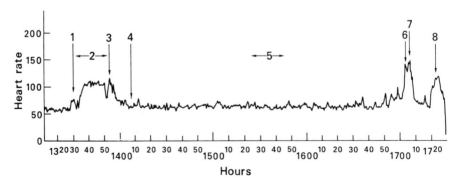

*Figure 13.1. Heart rate in coastal pilot: (1) dressing, (2) walking to the pier, (3) entering the boat,
(4) departure, (5) piloting, (6) climbing steps, (7) transferring to the pilot boat,
(8) arriving at the pilot's station (from Höyem-Johansen and Natvig, 1978).*

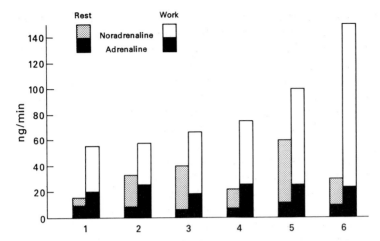

Figure 13.2. Urinary catecholamine elimination in different occupations: (1) coastal pilots, (2) shift workers in a steel factory, (3) chief operator in a cement factory, (4) fishermen, (5) assembly line operators, (6) War Academy cadets in simulated battle course (from Höyem-Johansen and Natvig, 1978).

urinary catecholamine elimination (Figure 13.2). The mean work load corresponded to about 20 per cent, and rarely exceeded 50 per cent of the heart rate reserve (HRR). The highest heart rates were recorded when leaving the ship, by climbing down a dangling rope ladder over the side of the ship and jumping into the small pilot boat which was moving along close to the ship (Figure 13.1).

Jungmann (1960) measured heart rate and body temperature in sailors in the tropics. Hampton (1974) has examined different criteria for the assessment of the ability to tolerate stress as part of the procedure for the selection of crew members. More recently, Bjørklund and Heien (1987) have prepared a manual for the selection of maritime crew members. Their elaborate selection system is based on past perform-ance, formal qualifications, language capability, specialized skills and qualifications related to seamanship, and a comprehensive test battery.

A group of German workers (Goethe *et al.*, 1978a), have computed a comprehensive descriptive survey of German sailors on board 24 different ships. The survey included vibration, noise, micro- and macro-environment, heat stress and lighting conditions at different work places on board. Some assessment of work loads were also carried out on the basis of recorded heart rate, oxygen uptake, body tempera-ture and urinary catecholamine elimination. The authors concluded that eight per cent of those examined showed evidence of mental stress, and that this kind of stress was most pronounced among the crew working on deck while in port, and among the engine room crew members while out at sea.

At the international symposium on human factors in the design and operation of ships, held in Gothenburg, Sweden in 1977, a great deal of emphasis was placed on the functional design of the bridge, and the use of bridge simulators (Andersson *et al.*, 1977).

A substantial part of the maritime literature deals with general health aspects of sailors. This is particularly the case with a series of publications from the Polish Institute of Maritime and Tropical Medicine in Gdynia (Tomaszewski *et al.* 1976; Filikowski, 1977, 1978, 1979; Ebert, 1979; Jaremin *et al.*, 1981; Rembowski, 1981; Tomaszewski *et al.*, 1981; Waskiewicz and Banaszkiewicz, 1981a, b; Dolmierski *et al.*, 1986; Kotlowski *et al.*, 1986). More specifically, Tomaszewski *et al.* (1976) reviewed the risk factors for coronary heart disease among seamen. On the basis of their observations of 43 seamen hospitalized due to coronary heart disease during the period 1966 to 1975, they concluded that the most frequent risk factor was cigarette smoking (in excess of 20 cigarettes per day over a period of 15–44 years). Other risk factors were overweight, high cholesterol levels and hypertension. Some of these conclusions were supported by Mundal *et al.* (1982) in their study of latent ischemic heart disease in sea captains. In their study of 110 apparently healthy Norwegian captains on ocean-going ships, a near maximal cycle ergometer test revealed a pathological exercise electrocardiogram for 10.0 per cent, while the corresponding figure for a comparable group of Oslo men was 4.6 per cent. For a comparable group of Norwegian coastal pilots, the figure was 11.8 per cent. The significant difference in the prevalence between sea captains and the Oslo men could not be explained by differences in serum lipids, blood pressure, or a family history of coronary heart disease. The captains were taller and more physically fit than the Oslo men, but they were significantly heavier and had a high prevalence of heavy smokers. Ten of the 11 captains with a pathological exercise electrocardiogram were, or had been, heavy smokers, i.e. smoking more than 20 cigarettes a day. In the Oslo men, 5.5 per cent smoked more than 20 cigarettes per day, as against 10.1 per cent in the case of the coastal pilots (Erikssen *et al.*, 1981b).

Tenfjord *et al.* (1983) examined 750 Norwegian seamen, 30–59 years old, with respect to coronary heart disease risk factors (smoking habits, serum cholesterol and blood pressure). The results show that the examined seamen, both officers and men, had a very high score compared with a control group from the normal Oslo population. Excessive tobacco smoking and increased serum cholesterol levels contributed largely to the risk profile.

Cook and Cashman (1982), observed a higher incidence of ectopic heart beats as judged by ambulatory ECG recordings in coastal pilots when they were involved in demanding activities, especially when maneuvering ships in hazardous situations, often associated with greatly increased heart rate.

White (1977) has presented a general description of the health standards and practices in the United States merchant marine. Barnes (1983) subjected a group of 153 randomly selected merchant marine officers to the Rorschach personality profile test. The results were compared with similar army data. Merchant marine officers had higher mental activity and organizational ability, but the thought process was 'disturbed owing to weakening of reality boundaries, thought disintegration, over-inclusive thinking, acute affectional anxiety and uncontrollable emotional reactions, when encountered by occupational hazards, psychosocial, sexual deprivation and frustrations'.

The popular opinion that the prolonged absence of seafaring fathers from home destroys their children's affection for them could not be confirmed by Rembowski (1981) in his study of the emotional relations in two-children Polish seaman families.

Hoiberg and McCaughey (1984) have drawn attention to the traumatic after effects, including neuroses, of collisions at sea. They examined the survivors from the guided missile cruiser USS Belknap over a three-year period, after the collision with the aircraft carrier USS Kennedy in the Mediterranean Sea on 22 November 1975. Significantly more of the Belknap survivors than a comparable control crew were hospitalized or separated from the service because of neuroses. Similar findings were reported by Dolmierski *et al.* (1986) in Polish seamen who were subject to a sudden and unexpected peace-time bomb attack in a port in Vietnam.

Collins *et al.* (1971) carried out a metabolic balance study in three crew members on an oil tanker when the ship was located in the Bay of Biscay, in the South Atlantic and in the Persian Gulf, in order to investigate possible changes in nutritional requirements, and the effect of acclimatization to heat on nutrient losses in sweat. They found that the intakes of water, sodium, potassium, nitrogen and iron provided by the ship's diet were adequate, or more than adequate. Sweat concentrations of sodium, potassium, and nitrogen were related to the rate of 24-hr sweat loss, but were not altered by heat acclimatization. Under the existing environmental and dietary conditions on board the daily sweat losses of potassium, nitrogen and iron had no significant effect on the electrolyte balance.

As part of their comprehensive research project on psychophysical stress of sailors on board modern ships, Goethe *et al.* (1978b) evaluated the noise levels on board 34 seagoing and 22 inland-waterway German ships. On the basis of their findings they put forward a series of recommendations as to future noise-measuring methods, maximal permissible noise levels, and recommendations as to what consequences their findings might have on the construction of new ships.

Szczepanski and Zaborski (1981) evaluated the efficiency of different types of hearing protectors for persons employed at various work stations on board merchant marine ships. They recommend that the selection of suitable hearing protectors should depend on the characteristics of the noise spectrum at a given work station, and should possess good damping properties, especially at the high frequency levels. They also pointed out that some of the commonly used hearing protectors are unsuitable due to perspiration in hot environments, and that engine room personnel should wear external protectors only, because the internal ones possess poor damping properties in all the frequency ranges.

Our own studies of stress at sea

It is generally agreed that the sailor's occupation at sea is associated with an exceptionally high accident rate. There may be a number of reasons for this. Conceivably the nature of the sailor's work on board, and the conditions under which it is carried out, may in themselves contribute to the high accident rate. This includes such factors as noise, visibility, cold, motion, positioning of equipment and machinery, slippery deck surfaces, work rhythms and shift work. The relative importance of any one of these factors might be assessed by a systematic investigation of individual seamen and their working conditions on board. This should include human factors, in particular factors associated with prolonged persistent mental stress, and their effect on alertness and the ability to concentrate. This would be of special interest in view of the observed association between personal conflicts and rates of accidents in fighter pilots, etc. (see Chapters 3 and 6).

On this background our Institute of Work Physiology in Oslo, at the request of the Norwegian Shipowners Association, decided to become engaged in a research project devoted to the safety at sea, during the period 1976 to 1980 (Rodahl, 1980). The objective was to contribute to the clarification of some of the above mentioned questions by a series of specific studies of individual crew members on board ships of the type which conceivably also would be operating in the future. These studies constitute an extension of previous studies carried out by the Institute on Norwegian fishermen (see Chapter 12).

Material and methods

Ninety-eight crew members of ten different ships were examined during ordinary sea voyages in different parts of the world (see Table 13.1). In all cases the crew was informed about the purpose of the study and the procedures to be used. As expected, a positive attitude of the ship's master turned out to be essential for the successful outcome of the study.

For studies of this kind, a state of complete mutual confidence

Table 13.1. Ships examined.

Name	Size (tonnage)	Type	Location	Crew	Number of subjects
Toyama	40 000	Container	Far East-Europe	31	30
Tender Pull	460	Tugboat	North Sea	9	6
Vindafjord	12 000	Cargo	East Africa-Europe	26	8
Bergensfjord	8 000	Cargo	North Atlantic	23	6
Texaco Bergen	18 000	Oil tanker	Europe	27	4
Tustna	675	Car ferry	Kristiansund	14	7
Harald Jarl	2 500	Coastal steamer	Bergen-Kirkenes	41	7
Tarn	12 700	Cargo	Persian Gulf-Europe	22	13
Tarago	20 655	Cargo	Persian Gulf-Europe	28	6
Docecanyon	273 000	Combination oil iron ore	Brazil-Japan	32	11

between the research team and the subject is essential. The study involved prolonged recordings of different parameters, usually during a complete 24-hour period, sometimes longer. In fact, in one of the studies, 24-hour rectal temperatures were recorded over a period of several weeks in two of the subjects. Studies of this kind are extremely time-consuming. For this reason it is not possible to collect a vast amount of data, nor can one expect to be able to study a large number of subjects. Concentrating on a smaller number of subjects may, on the other hand, provide a more thorough assessment, and since our experiences so far have indicated a high degree of reproducibility of the observations made (see Figure 13.3), they may appear to be fairly representative of the type of problem under study, in spite of the limited number of subjects.

In the course of this study it soon became evident that seamen tend to be excellent subjects. They are dependable, punctual, observant, co-operative and easy to get along with. They had a positive attitude to the study and, even more important, they were available at all times since there was nowhere else to go as long as the ship was in open water.

In each case, a separate laboratory was established in one of the cabins on board. As a rule, at least two subjects were studied simultaneously. The duration of the stay on board each ship varied from one to more than four weeks. The study included a general survey of the individual work places, supplemented by personal interviews with those working there. Occasionally questionnaires were used to clarify specific points of interest. A detailed time-activity log was kept for each subject, covering his activities, minute by minute. As a rule this was done by the subject himself, checked and supplemented by the observer.

An assessment of the subject's reaction to his work stress was made by the continuous recording of his heart rate by battery-operated

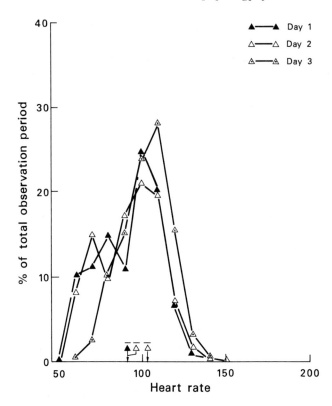

Figure 13.3. Heart rate in cabin attendant onboard M/S Tarn *on three different days (from Maehlum et al., 1978).*

miniature recorders. By establishing the relationship between heart rate and work load for each subject individually, with the aid of a cycle ergometer, the recorded heart rate could be translated into actual work load. In the case of one ship (M/S *Toyama*), the subjects' maximal physical work capacity (aerobic power) was assessed with the aid of a mechanically-braked Monark cycle ergometer on board the ship.

In some cases it was possible to record the blood pressure of the subject (the captain) during critical work situations, with the aid of a Bosomat II automatically inflatable blood pressure cuff, and a 4-meter long tube, enabling the subject to move about on the bridge and to perform his job without undue restriction.

As an indication of the general stress reaction, the urinary elimination of catecholamines (epinephrine and norepinephrine) were determined in urine samples collected on board, and later analyzed fluorometrically in the laboratory at the Insitute by the method of Andersson *et al.* (1974). The assessment of the subject's heat strain was done by continuously recording the skin and rectal temperatures, usually over a 24-hour period, with the aid of sensors connected to the Medilog recorder.

In an attempt to assess the physiological reaction on the changing of the ship's time when travelling from west to east or from east to west, crossing different time zones, physiological functions such as heart rate and body temperatures, which are affected by circadian rhythms, were continuously recorded in the same person for extended periods.

Results

One of the most extensive studies in this series was the initial study on board the container vessel M/S *Toyama*, extending over more than four weeks, en route from Singapore to Panama via Japan. Of a crew of 31, 30 volunteered to participate. It turned out that the main results from this first study, in a series of ten ships, were essentially verified or substantiated by the subsequent studies on the other nine ships.

Physical work capacity

The physical fitness of the sailors, in spite of limited facilities for physical activity and training on board, appears not to be greatly different from that of comparable age groups in the population as a whole. The mean value for the estimated maximal oxygen uptake, based on submaximal heart rate, for all the 30 subjects tested on board the M/S *Toyama*, was 3.2 l O_2 per min, corresponding to 44 ml O_2 per kg bodyweight in the male crew members (Table 13.2). This is almost the same as the mean value for young Norwegian military recruits (about 45 ml per kg bodyweight). However, the scatter of the data was very large (21 to 58 ml per kg). The mean maximal oxygen uptake of the officers was 39 ml per kg as against 48 ml per kg for the rest of the crew (Rodahl *et al.*, 1977). This may partly be explained by the fact that the ordinary crew members were, on the whole, younger than the officers, and some of them were particularly active; three of them trained regularly during their time off.

Considerably lower values were found in the six crew members on the small tugboat T/B *Tender Pull*. Here the mean predicted maximal oxygen uptake was only $2.8 \, l \cdot min^{-1}$ (2.2–4.0) or $34 \, ml \cdot kg^{-1}$ bodyweight (Rodahl and Vartdal, 1977).

Physical work stress

The physical work load, in general, as assessed by the continuously recorded heart rate, is, on the whole, quite comparable to what is found in ordinary Norwegian industrial workers ashore (Figure 13.4). Individual values of the working heart rates in 23 crew members on

Table 13.2. *Physical characteristics of the subjects, M/S Toyama.*

Subject no.	Occupation	Age years	Height cm	Weight kg	Calculated Max $\dot{V}O_2$ (1/min)	Max $\dot{V}O_2$ (ml/min/kg)
1	Captain	52	176	74.5	2.3	31
2	Chief engineer	51	176	88.5	1.9	21
3	First engineer	32	188	68.5	2.5	36
4	Second engineer	32	174	72.0	3.0	42
5	Second engineer	46	191	92.5	3.5	38
6	Second engineer	24	191	86.0	4.1	48
7	Motorman	22	184	101.0	3.6	36
8	Motorman	20	191	83.0	3.6	43
9	Motorman	45	172	72.5	2.4	33
10	Steward	35	177	73.0	4.2	58
11	Cook	28	176	68.0	3.1	46
12	♀ Cabin attendant	23	172	66.0	2.2	33
13	♀ Cabin attendant	34	167	59.5	1.9	32
14	Deck hand	26	180	68.0	3.6	53
15	Deck hand	27	165	72.5	2.6	36
16	Deck hand	26	177	71.5	3.6	50
17	Deck hand	33	172	59.0	3.4	58
18	Repair man	30	184	68.0	3.3	49
19	Repair man	21	173	63.0	3.2	51
20	Apprentice	18	183	63.0	3.5	56
21	Apprentice	19	180	74.5	3.6	48
22	Radio operator	46	163	57.5	2.2	38
23	First officer	32	189	81.0	2.8	35
24	Chief officer	37	168	74.0	3.7	50
25	First officer	27	189	96.0	3.1	32
26	Foreman	52	—	50.0	—	—
27	Apprentice	22	171	63.5	3.1	49
28	Apprentice	22	185	69.0	3.4	49
29	Motor man	23	184	72.5	4.3	59
30	♀ Cabin attendant	55	154	63.0	1.6	25
	Mean	33	178	72.4	3.2 (26♂) 1.9 (3♀)	44.0 (26♂) 30.0 (3♀)

Note. Subjects nos. 10, 16 and 20 were engaged in regular physical training.

board the M/S *Toyama* are presented in Table 13.3. The heaviest physical work load, relatively speaking, is observed in women working in the mess and cleaning the cabins (Figure 13.5). For extended periods they might work at a load corresponding to 30–50 per cent of their maximal physical work capacity. These women, as is the case of some of our industrial workers ashore, tend to work continuously for protracted periods, and then to take prolonged rest periods, rather than doing shorter periods of work, interspersed with brief rest pauses. The heavy work load of the cabin attendant was a typical finding on all ships

Table 13.3 Mean work loads in crew members on board M/S Toyama

Subject no	Occupation	Mean heart rate				Work load while working (% HRR)	% of work period with HRR greater than 130
		Entire observ. period	Work	Free	Sleep		
1	Captain	78	87	82	64	22	0.0
2	Chief engineer	77	88	80	66	23	0.1
3	First engineer	85	86	–	78	15	0.0
4	Second engineer	79	90	72	63	23	1.2
5	Second engineer	78	93	81	62	29	0.4
6	Second engineer	93	96	71	–	27	0.0
7	Motorman	73	87	72	63	18	0.5
8	Motorman	73	102	74	63	33	0.7
9	Motorman	90	104	83	81	32	2.0
10	Steward	70	78	74	58	17	2.7
11	Cook	85	86	81	–	12	0.2
12	Cabin attendant	79	97	78	62	27	0.2
13	Cabin attendant	80	99	75	61	29	2.1
14	Deck hand	77	92	77	58	26	0.8
15	Deck hand	66	77	64	53	17	0.0
16	Deck hand	78	84	88	64	15	0.4
17	Deck hand	82	102	74	62	33	0.3
18	Repair man	77	92	70	64	26	0.3
19	Repair man	94	107	95	79	30	3.7
20	Apprentice	79	95	85	53	28	0.0
21	Apprentice	91	110	93	73	32	0.6
22	Radio operator	71	75	73	65	11	0.8
23	First officer	86	–	90	80	–	–
	Mean	80	92	79	65	24	0.8

examined. Another example of this is presented in Figure 13.6. In contrast to this, the cook on board the M/S *Toyama* enjoyed an exceptionally light work load, as is evident from Figure 13.7.

Next to the cabin attendants, it is the crew working in the engine room that is regularly subjected to the greatest physical work load. The lower the rank, the greater the physical work load. Climbing up and down steps and ladders constituted the greatest physical work stress (Figure 13.8). The captain, as a rule, is subject to very little physical stress, as is also the case of the chief engineer, the wireless operator, and on some ships also the cook (see Figure 13.7).

It should be noted, however, that on board small ships, such as the tugboat T/B *Tender Pull*, operating at the oil fields in the North Sea, the almost constant motion due to waves and wind caused a more or less constant elevation of the heart rate (and presumably also of the energy expenditure) even when resting and during sleep, because of the

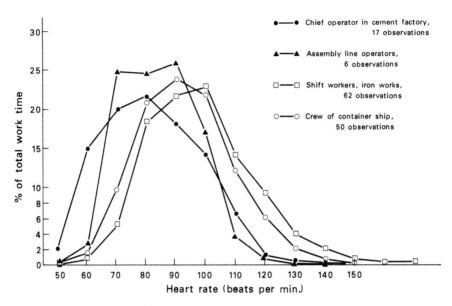

Figure 13.4. Work loads in different occupations, expressed in terms of heart rate recorded during an ordinary work day.

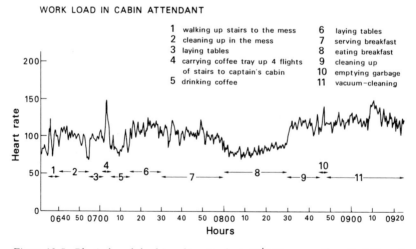

Figure 13.5. Physical work load in cabin attendant, M/S Toyama (from Rodahl et al., 1977).

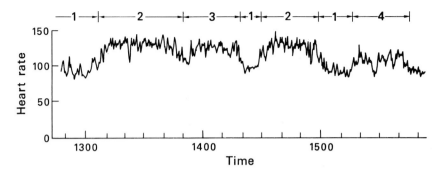

Figure 13.6. *Physical work load in cabin attendant M/S* Bergensfjord: *(1) break, (2) washing floor, (3) polishing floor, (4) odd jobs (from Rodahl and Huser, 1978).*

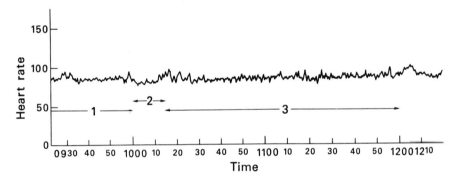

Figure 13.7. *Physical work load in the cook, M/S* Toyama: *(1) odd jobs in the kitchen, (2) coffee break, (3) preparing lunch (from Rodahl, et al., 1977).*

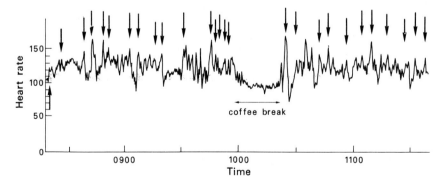

Figure 13.8. *Physical work load, as evidenced by heart rate, in a motorman onboard M/S* Vindafjord. *Arrows indicate ladder climbing (Blix et al., 1979).*

Table 13.4 *Mean heart rate in crew of T/B Tender Pull.*

Subjects	Entire ob-servation period	Anchor handling	Duties without anchor handling	Free	Sleep	Engine room duty	Other duties
Captain	96 (50–165)	115 (85–165)	94 (50–133)	90 (68–129)	78 (63–94)		
Chief engineer	89 (57–153)	97 (77–143)	118 (64–148)	91 (67–153)	73 (57–129)		
Engineer	84 (52–135)			90 (68–128)	69 (52–122)	104 (69–135)	
Deck hand no. 1	80 (55–177)	99 (75–180)	73 (56–116)	76 (59–122)	68 (59–89)		
Deck hand no. 2	87 (61–162)	102 (73–159)	84 (67–113)	86 (67–110)	78 (61–110)		98 (75–162)
Deck hand no. 3	84 (67–127)		82 (67–127)	88 (70–123)	77 (68–77)		
Mean	87	103	90	87	74	104	–

necessity of counteracting the motion of the ship (see Tables 13.3 and 13.4). The mean heart rate during the time off duty was 79 on board the large vessel M/S *Toyama*, as against 87 on board the smaller vessel T/B *Tender Pull*. During sleep, the mean heart rate was 65 on board M/S *Toyama*, as against 74 on board T/B *Tender Pull*.

The work of the deck hands out at sea may, on the whole, be rather uneventful and the work load moderate with occasional periods of

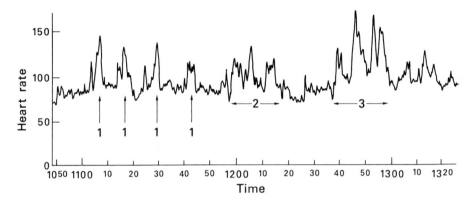

Figure 13.9. *Work load of a deck hand on board the small tugboat T/B* Tender Pull *during anchor handling at the offshore oil fields in the North Sea: (1) anchor handling, (2 and 3) handling the buoy in heavy seas, while the waves are breaking in over the deck (from Rodahl and Vartdal, 1977).*

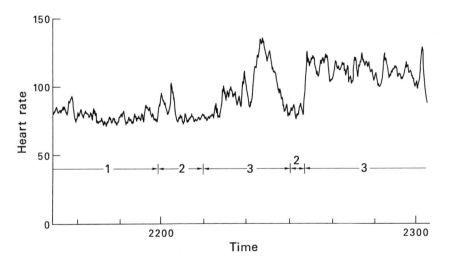

Figure 13.10. *Work load of deck hand, on board coastal steamer M/S* Harald Jarl; *(1) standing at the wheel, (2) break, (3) securing cargo on deck in heavy seas (from Rodahl et al., 1980a).*

extreme work stress. An example of this is the deck hand on board the tug boat T/B *Tender Pull* during anchor handling (Figure 13.9), and one of the deck hands on board a coastal steamer during a sudden episode of rough weather when it became necessary to secure the deck cargo in a hurry (Figure 13.10).

The work of the ships' officers, when sailing in open waters, is

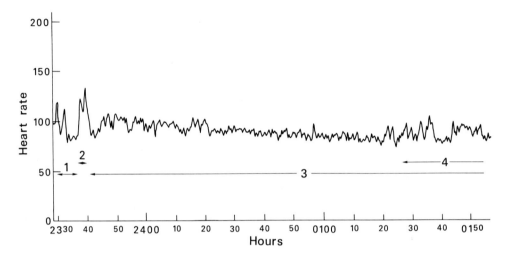

Figure 13.11. *Heart rate in the second officer on night watch on the bridge on board M/S Toyama. The falling heart rate is due to the circadian rhythm: (1) in the cabin, (2) walking up to the bridge, (3) standing watch on the bridge, (4) worried about problems at home (from Rodahl et al., 1977).*

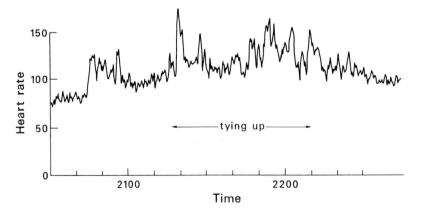

Figure 13.12. *An example of the heavy physical work stress in the 1st mate on board M/S Bergensfjord when tying up the ship in port (from Rodahl and Huser, 1978).*

monotonous, without much physical effort (Figure 13.11). During arrival or departure from ports, however, they may encounter rather severe physical work loads for brief periods (Figure 13.12). It appears that the deck officers were exposed to excessive work stress during loading and unloading operations in the Persian Gulf, where heat stress is also a problem of considerable magnitude (Figure 13.13).

Sleep deprivation may be another serious problem for the chief officer (first mate) on combination ships (such as oil and iron-ore carriers) as in the case of the S/S *Docecanyon* during the unloading of oil, immediately followed by tank cleaning in preparation for the loading of ore. In such cases there may be very little opportunity for sleep over a period of several days. In the case of the S/S *Docecanyon*, the chief officer had only a couple of hours sleep in three days running.

When it comes to mental stress, the captain is particularly exposed. Manoeuvering in and out of harbours, as a rule, causes an elevation of the captain's heart rate (Figures 13.14; 13.15; 13.16 and 13.17) and blood pressure (Figures 6.1, 13.18 and 13.19) and in some cases a measurable increase in catecholamine elimination (Figure 13.20). This occurs even though the captain, standing on the bridge, is not involved in any physical effort which might explain some of the observed changes. They must therefore be attributable to mental tension or stress. The elevated heart rate in the captain on the bridge during the

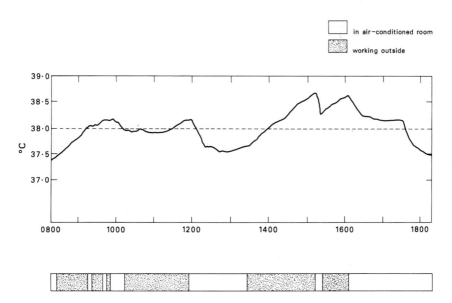

Figure 13.13. Rectal temperature in chief officer of a cargo ship, (M/S Tarn*) in the Persian Gulf in the course of a working day (Maehlum et al., 1978).*

manoeuvering in and out of harbour is a fairly constant phenomenon in all the captains of ocean-going vessels which we have examined. The fact that this also occurred in the captain on board the tugboat T/B *Tender Pull* when sitting at the controls on the bridge, manoeuvering his ship close to the side of the huge crane vessel in very rough weather, when the danger of deck hands being hurt was considerable, indicates that the stress is of a psychological nature (Figure 13.17).

Our findings suggest that the intermittent stress, to which captains of ocean-going ships are exposed, is worse than the more or less continuous stress encountered in frequent calls at ports, to which one more or

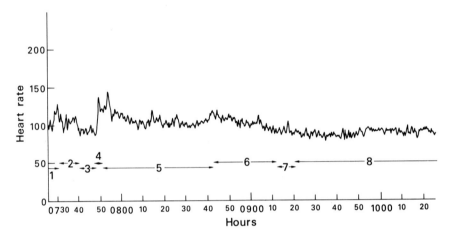

Figure 13.14. Evidence of mental tension in the captain of M/S Toyama *during the departure from Kobe harbour in Japan: (4) walking up to the bridge, (5) standing on the bridge with coastal pilot, (6) in charge on the bridge after the coastal pilot had left the ship, (8) standing on the bridge in open water (from Rodahl et al., 1977).*

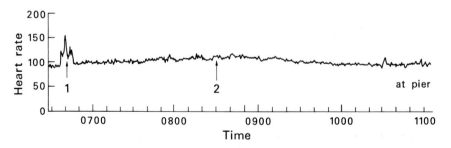

Figure 13.15. Evidence of mental stress in captain on board the supertanker S/S Docecanyon *arriving at Angra Dos Reis: (1) in the cabin, (2) ordering all visitors to leave the bridge (from Rodahl et al., 1980b).*

less becomes accustomed. An example of this is the captain on board the 18,000-ton product carrier M/T *Texaco Bergen*, with a crew of 27, distributing oil from the port of Brofjorden in Sweden to a number of European ports, involving frequent calls at ports with brief sailing periods in between (Rodahl *et al.*, 1979). From Figure 13.21 it appears

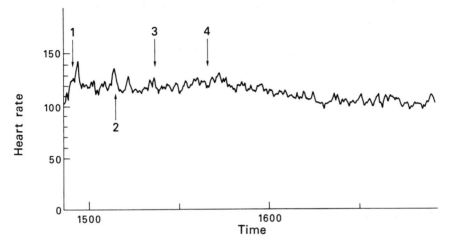

Figure 13.16. *Heart rate reaction of captain on board M/S* Vindafjord *during the departure from Le Havre during unfavourable weather conditions: (1) coastal pilot arriving, (2) discussion with chief officer, (3) leaving the port, (4) the coastal pilot departing (from Blix* et al., *1979).*

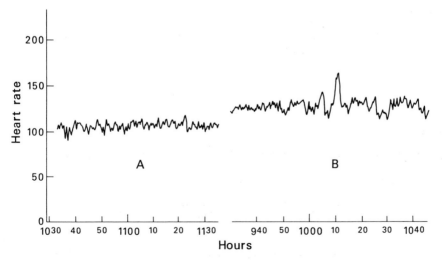

Figure 13.17. *Heart rate in captain on board tug boat during anchor handling; (A) favourable conditions, (B) difficult conditions. (from Rodahl and Vartdal, 1977).*

that the captain's heart rate reactions to the manoeuvering in and out of port is minimal (in this case the blood pressure rise is also moderate, see Table 13.5). Figure 13.22 shows the heart rate of the captain of a small local ferry, constantly moving in and out of port the year round in a limited coastal area on the west cost of Norway.

On the basis of these observations one might have expected the heart rate of the captains of the coastal steamers constantly navigating up and down the Norwegian coast from Bergen to Kirkenes and back the year round to be less affected by the manoeuvering in and out of port. This, however, did not always turn out to be the case. The explanation is probably that the weather and current conditions, in addition to visibility etc., are extremely variable, causing unexpected incidents and conditions.

As already pointed out, it appears reasonable to assume that the elevated heart rate in the captains during manoeuvering in and out of port is due to mental tension caused by the responsibility which they carry, since they are not doing any physical work to speak of. This is supported by the observation of the heart rates of the captain (Figure 13.23), and his chief officer (Figure 13.24), on board M/S *Bergensfjord* during the approach to Copenhagen (Rodahl and Huser, 1978). They were both on the bridge but the captain, who was in charge, reacted

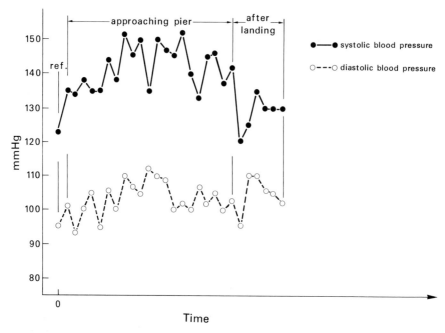

Figure 13.18. Blood pressure in captain of coastal steamer approaching the port of Hammerfest.

with an increased heart rate, while the chief officer did not (Figures 13.23, 13.24).

As previously mentioned, the manoeuvering in and out of ports was also associated with elevated blood pressures. The greater the heart rate increase, the greater the blood pressure elevation. Furthermore, urinary

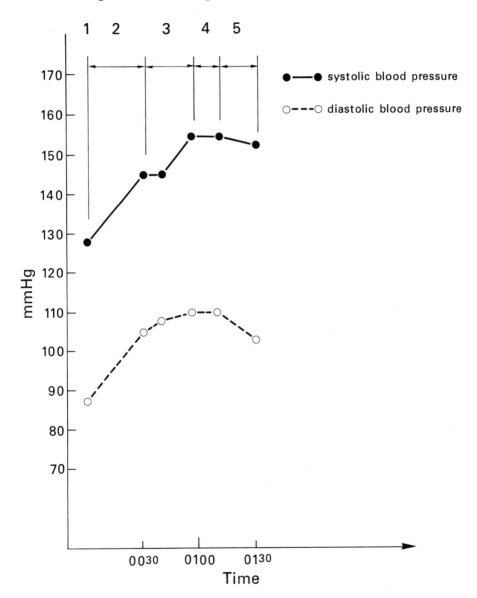

Figure 13.19. Blood pressure in captain of supertanker during departure from Angra Dos Reis:
(1) control values, (2) on the bridge, (3) waiting for the arrival of the tug boat,
(4) leaving the port, (5) full speed ahead.

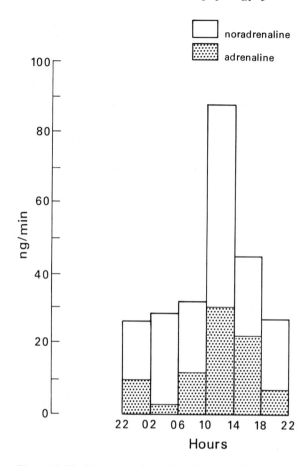

Figure 13.20. Urinary catecholamine elimination of captain on board M/S Vindafjord *during the departure from Le Havre (from Blix et al., 1979).*

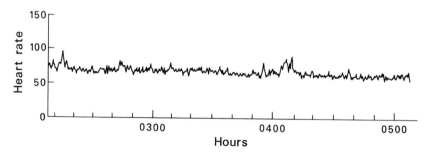

Figure 13.21. Heart rate of the captain on board the product carrier M/T Texaco Bergen, *arriving at Nörre Sundby in Denmark early in the morning, from the time the coastal pilot enters the ship at 0205 hours, until the ship is tied up at the pier at 0520. The heart rate is fairly steady, around 70–80 beats per min (from Rodahl et al., 1979).*

catecholamine elimination, especially epinephrine, also mirrored the captain's mental tension, associated with the stress of bringing his ship safely in and out of port (Figure 13.20). This was clearly the case in all the five ships where urinary catecholamine eliminations were assayed. It should be pointed out, however, that the catecholamine assay did not show any evidence of strain which was not revealed by the recorded heart rates.

From the observations made during this study it is evident that the captain is subject to considerable mental strain. Evidently, the existing rules and regulations impose an unreasonable mental burden on the

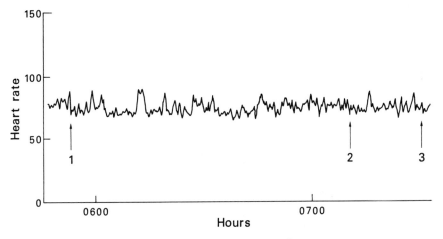

Figure 13.22. Heart rate in the captain of the local ferry B/F Tustna: (1) arrival at Forsnes, (2) arrival at Kyrhaug, (3) arrival at Straumen.

Table 13.5. Systolic and diastolic blood pressure (mm Hg) in the Captain of M/S Texaco Bergen.

	Systolic	Diastolic
Control value	123 ± 4.9	92 ± 4.3
	1.7 (8)	1.5 (8)
Leaving Brofjorden	128 ± 5.6	100 ± 5.2
	1.1 (27)	1.0 (27)
Arriving Fredricia	129 ± 7.9	96 ± 9.6
	3.0 (7)	3.6 (7)
Leaving Fredricia	123 ± 5.2	82 ± 4.1
	1.4 (14)	1.1 (14)
Arriving Nörre Sundby	112 ± 6.6	88 ± 6.2
	1.9 (12)	1.8 (12)

Note. Figures denote mean, standard deviation, standard error of the mean, and number of observations (in parentheses).

captain, who feels obligated to be on the bridge in all instances when difficulties may be encountered. In this respect there may be something to be learned from the arrangements practiced on board some of our coastal steamers, with two fully competent shipmasters taking turns on the bridge, or from civil aviation, with two competent pilots in the cockpit.

It is conceivable that the age of the captain, his level of competence, and his emotional make-up also play a role, and that obviously some individuals are better suited than others to become captains. In any case, it is evident that leadership training is an important part of the education of captains. It is the captain who sets the tone on board his ship. He has it in his power to make his ship a pleasant or an unpleasant

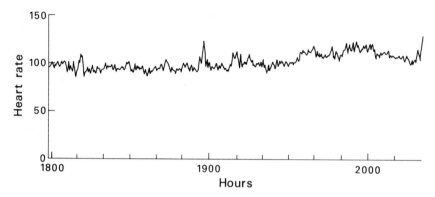

Figure 13.23. *Heart rate of the captain onboard M/S* Bergensfjord, *during the approach to Copenhagen (from Rodahl and Huser, 1978).*

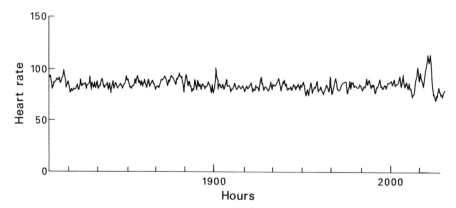

Figure 13.24. *Heart rate of the chief officer on board M/S* Bergensfjord *during the same approach to Copenhagen as in Figure 13.23 (from Rodahl and Huser, 1978).*

place of work. Our observations also indicate all too often that there is too great a gap between the captain and his officers in terms of competence and responsibility. In fact our observations indicate that the state of professional competence at all levels of our sailors is a question of utmost importance for safety at sea.

The present system, making the shipmaster responsible for all matters at all times, occasionally causes him to obtain insufficient sleep and rest to function properly. This is especially the case in heavy traffic, bad weather and when sailing during the night, and when loading or unloading at a port during the day several days in a row.

Assessment of the level of heat stress was made on board two ships (the M/S *Tarn* and the M/S *Tarago*) in the Persian Gulf, and on board the S/S *Docecanyon en route* to South America. In the latter case, Botsball temperatures recorded in the engine room did not show excessively high values, and rectal temperatures in excess of 38 °C were rarely observed in members of the engine room crew (Rodahl *et al.*, 1980b).

A study by Maehlum *et al.* (1978) of the crew on board the M/S *Tarn* in the Persian Gulf during the month of September showed that in the case of the chief officer, the rectal temperature exceeded 38 °C in 40 per cent of the entire work period (Figure 13.13). In the deck hands examined, the rectal temperature exceeded 38 °C for 13 per cent of the time they worked in the open air on deck. In the officers, working outside on deck, the heart rate exceeded 130 beats per min (corresponding to about 50 per cent of their maximal work capacity) for more than 50 per cent of the time, as against only about 9 per cent for the deck hands on the average. In the engineers, the heart rate exceeded 130 beats per min in 18 per cent of the time they worked in the engine room. The recording of the bodyweight before and after the work shift, combined with the recording of fluid intake, showed that the fluid intake was ample to maintain the fluid balance, avoiding any evidence of hypohydration.

A subsequent study by Jebens and Bolling (1980) during the month of August on board the M/S *Tarago*, during a voyage in the Red Sea and the Persian Gulf, revealed considerable heat stress both on deck and in the engine room. The Botsball readings exceeded the recommended ceiling of 28 °C for extended periods, both in the engine room and on deck. The heart rate recordings, however, did not reach values comparable to those observed by Maehlum *et al.* (1978). In the M/S *Tarago* engineers, the heart rates exceeded 130 beats per min in only 5 per cent of the time they worked in the engine room, and their rectal temperature exceeded 38 °C in only 3 per cent of the time.

It is thus evident that the heat stress in the engine room in ships in the tropics may not necessarily be any greater than that encountered on deck. Furthermore, the use of modern air-conditioning systems may greatly serve to eliminate problems related to heat stress. At any rate, adequate fluid intake remains an important measure of preventing the pathophysiological consequences of heat stress in sailors in the tropics.

The effects of crossing time zones

The initial studies on board the M/S *Toyama* and the M/S *Bergensfjord* indicated that a fair percentage of the crew experienced different degrees of untoward effects from having to set the clock back or forth when crossing different time zones.

During a voyage across the Atlantic from New York to Copenhagen on board the M/S *Bergensfjord*, the 24-hour recording of the heart rate and body temperature was repeated on three different dates in one subject. During this voyage the clock was put forward four hours in the course of five days. The results indicate that, in spite of a time-shift of four hours, the basal circadian rhythms were surprisingly well maintained, and appeared largely to be keyed to the activity pattern.

This problem was further pursued during a study on board the S/S *Docecanyon en route* from Cape Town in South Africa to Angra Dos Reis and Tubarao in Brazil. During this east-west voyage, the ship's clock was put back one hour every other night. The rectal temperature was continuously recorded every other 24-hour period from October 27 to November 11 in two subjects: the first mate who worked 4-hour shifts, and 8 hours free; and the chief engineer who only worked during

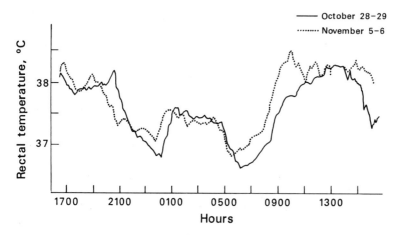

Figure 13.25. Rectal temperature (24-hour) of the first officer on board the S/S Docecanyon en route from South Africa to Brazil, recorded at the beginning and at the end of the voyage. The ship's clock was put back one hour every other night. The record may indicate a slight shift in the circadian rhythm to the left (from Rodahl et al., 1980b).

the day. In the officer working regular shifts there appeared to be a slight shift in the circadian rhythm to the left, when comparing the temperature record for the first and the last day of the voyage (Figure 13.25), suggesting a possible lag in the rhythm. This lag disappeared during the subsequent days when the ship remained at the same latitude on the east coast of Brazil. No such phase shift was observed in the case of the chief engineer who worked only during the day and was free to sleep regularly during the entire night. None of these subjects experienced any sleeping difficulties during the study.

A general survey based on questionnaires and personal interviews of the entire crew revealed that problems connected with changing the ship's time is less pronounced when travelling from east to west, than is the case when travelling from west to east, especially when the changing of the time is made during the night. Of a crew of 32 (including four women), 14 (43 per cent) complained of problems connected with the change of the local time. Of these, 11 stated that their problems occurred only when travelling from west to east (changing the time one hour ahead), two when travelling from east to west (changing the time one hour back), and one claimed to have problems both ways. The complaints included fatigue, difficulty in falling asleep, disturbed or insufficient sleep and gastro-intestinal complaints.

The use of ship simulators

The role of aircraft simulators in the training, retraining and certification of aviators is well established. This is a fact, in spite of the concern by some of those professionally involved, about some of the current activities in developing and selling simulators (Hopkins, 1975). The introduction of simulation techniques in the marine world is a recent development (Paffett, 1978).

Since the purpose of a simulator is to simulate reality, the value of any simulator must obviously depend on how well it simulates the happenings which it is trying to mimic. In a preliminary study at the Maritime Bridge Simulator in Trondheim, Norway, a coastal pilot and two students of a maritime college were tested. Their reactions during simulated ship maneuvering procedures are shown in Figures 13.26, 13.27 and 13.28. The coastal pilot showed evidence of considerable tension prior to the test, but appeared perfectly calm during the actual maneuvering, except when he was suddenly asked to make a complete turn in the middle of the Channel, awaiting further instructions. A final rise in heart rate occurred while he anxiously awaited the score of his test (Figure 13.26). One of the students exhibited a perfectly reasonable

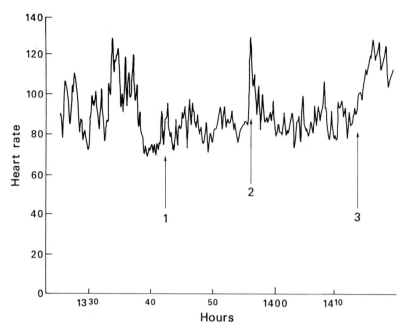

Figure 13.26. Heart rate in an experienced coastal pilot during a simulated maneuver into the port of Southampton: (1) start, (2) turning maneuver, (3) stop.

reaction to the challenge of the test (Figure 13.28), while the other student appeared completely out of control (Figure 13.27).

In a subsequent study, an attempt was made to compare the shipmaster's reaction in terms of heart rate response and professional performance during the maneuvering into specific ports in real life, compared to the simulator approach to the same ports (Wöhni *et al.*, unpublished data). The purpose of the project was to find out to what extent the captain's physiological reactions on the bridge simulator corresponded to his reactions in real life. The study included five sea captains (one of them being studied on two occasions), calling at Europort in the Netherlands, and Rafnes in Norway. In the first case, the research team was landed by helicopter together with the coastal pilot on board the ship out in the Channel before the approach to Europort. In the latter case the team boarded the ship at Rafnes, and followed it across the North Sea to Teeside in England, and back to Rafnes.

The captain's reactions were judged by his continuously recorded heart rate as an indication of autonomous nervous system reaction, paired with time-activity records. The actual arrival at port took place during the night, and the bridge simulations were made under night conditions. The reactions observed at the simulator were, on the whole,

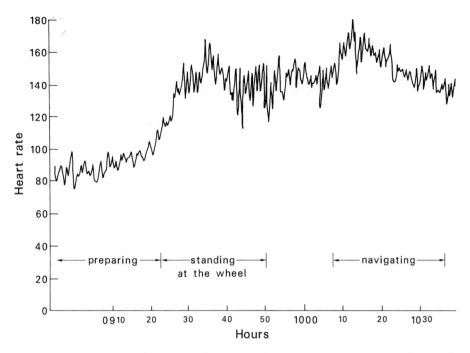

Figure 13.27. *Heart rate of inexperienced maritime college student maneuvering into the port of Rafnes. His heart rate is greatly elevated both when he is in charge of the manoeuvring and when he is standing at the wheel.*

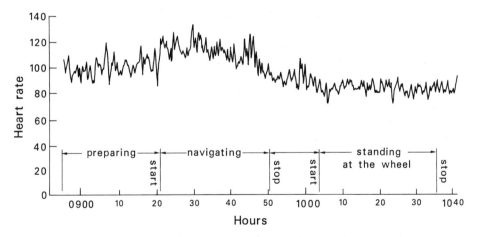

Figure 13.28. *Heart rate of another inexperienced maritime college student, elevated when in charge of the maneuvering of the ship but not when he is standing at the wheel taking orders.*

quite similar to those observed in real life. This was in agreement with the captain's own subjective impressions.

Bjørklund *et al.* (1987) have assessed the effect of maritime simulators as a training device for experienced shipmasters. Twelve captains performed three simulator exercises each, while their heart rate and physical activity were recorded. Mental load was assessed by a psychological test. Four port areas were used in the exercises: Ras Tannurah (Saudi Arabia); Europort (Netherlands), Solent (Southampton, England), and Rafnes (Norway). Each exercise was plotted on a map, and used for analysis in terms of the ship's heading and speed. Ship groundings were plotted, based on depth indications. Each simulator exercise lasted from 25 to 90 minutes, depending on the maneuvering area, ship speed and especially groundings, which terminated the exercise.

Six of the captains exhibited a high-level of maneuvering competence, three of them moderate level, and three were evaluated as exhibiting a low level of competence. The physical activity logs indicated that the captains who started the exercise without meticulous planning were unable to complete the required turn operation successfully. Five captains performed all their three exercises without any accidents, three had one exercise with grounding, two had two exercises with groundings, and two grounded in all their exercises.

The effect of training was evident in the captains with high competence levels, but more so in the case of the captains with moderate competence levels. The captains with a low level of competence did not improve their performance during the simulator-based training at all.

In contrast to the flight simulator performance, where the pilot is immediately excluded from further flight duties after accidents in the simulator, the performance in the maritime simulator has no consequence for the sea captain, even if he repeatedly fails in his exercises. Furthermore, while aircraft pilots have to prove their competence twice a year, the ship-masters are granted their operating privileges without having to demonstrate their competence by tests repeated at regular intervals.

Chapter 14
Muscle tension

Introduction

As repeatedly pointed out in the preceding text, the human neuromuscular system is not ideally suited for static work, requiring prolonged, exerted isometric muscle contraction. Consequently, a large number of individuals are unable to tolerate intense static industrial work as a regular occupation for the entire working shift, day after day, for years on end, without ill effects. These effects may include muscular stiffness, pain, tension, or even symptoms or signs of neuromuscular disorders. The remarkable fact, however, is that some individuals do not develop any such ill effects, in spite of years of exposure to static work of the kind referred to above. It has been repeatedly observed that two individuals may be working side by side, doing exactly the same job, and one of them may develop neuromuscular complaints leading to prolonged sick absence, while the other does not. It is not merely a matter of an obvious faulty ergonomic working position or working procedure, for ergonomic improvements in many cases do not necessarily make any difference in terms of preventing neuromuscular disorders and sick absence due to them (Westgaard *et al.*, 1986). Evidently, sex and age are factors of importance. It appears that the work rate and work intensity (i.e. productivity) play a role. It is also conceivable that individual differences in the way the specific tasks are performed, including visual and neuromuscular concentration, arm support and the application of micropauses, may play a role. Similarly, it is conceivable that individual differences in mental stress affecting peripheral neuromuscular tension may play a role. This may, in turn, be caused by unsolved personal problems or conflicts. Or it may be caused by the inability to relax while involved in a monotonous manual static task, allowing the mind to be free, and the thoughts to drift on their own accord. The individual difference in the ability to sense the state of tension of one's own muscles (muscle sense), may also be a factor to be considered as a contributory cause of neuromuscular complaints as a consequence of static work.

In static muscular work, the maximal voluntary contraction (MVC) can be maintained for only a few seconds. A force representing 50 per cent of MVC can be maintained for about one minute. Forces less than 15 per cent MVC may, according to Rohmert (1968) and others (Figure 14.3), be held from 10 min up to hours (Simonsen and Lind, 1971). More recent data indicate, however, that the upper limit for isometric contraction maintained 'indefinitely' is well below 10 per cent MVC (Bjørksten and Johnsson, 1977; Westgaard et al., 1984). The endurance time depends to some extent on the muscle groups studied (such as anti-gravity muscles versus upper extremity muscles), fiber types, and individual differences in maximal muscle strength, as well as individual differences in motivation. The length of time which successive, isometric contractions can be maintained also depends on the length of the interval between each contraction. According to Funderburk et al. (1974) the rate of recovery of isometric endurance is higher the greater the tension.

Assessment of muscle tension

When discussing the causative relationship between static muscle tension and neuromuscular complaints, objective data on the magnitude of muscle tension is essential, not only in laboratory subjects, but also in individuals doing static muscular work in real life.

As we have seen in the preceding text, oxygen uptake and heart rate may be used to assess the magnitude of muscular work in general. But these parameters may not necessarily be suitable for the monitoring of localized muscular loads (Malmqvist et al., 1981). However, under certain conditions, muscular engagement, even of small muscle groups, will affect the heart rate, which may also respond noticeably to postural changes (Hanson and Jones, 1970). According to Funderburk et al. (1974) the heart rate increase may be directly related to the isometric muscle tension: the higher the tension, the greater the heart rate increase.

The contraction of the muscle fibers in a muscle is associated with small electrical changes, or oscillations, which can be picked up by suitable surface electrodes, placed on the skin over the muscle in question. The instrument picking up the electrical oscillations is known as an electromyograph, and the display of the myoelectrical oscillations is called an electromyogram (EMG). A simple example of such a direct electromyographic recording is presented in Figure 14.1, which shows that the greater the load on the muscle, the larger the oscillations. In more sophisticated quantitative electromyography the generated electrical activity is measured by rectifying and integrating the action potentials recorded from surface electrodes. These integrated action potentials are linearly related to the muscular force which is being exerted (Bigland and Lippold, 1954).

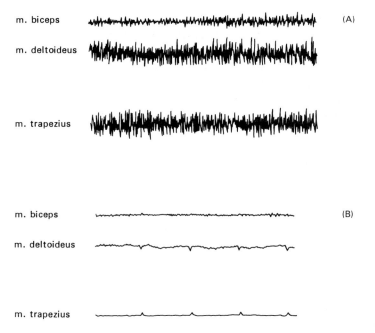

m. biceps (A)

m. deltoideus

m. trapezius

m. biceps (B)

m. deltoideus

m. trapezius

Figure 14.1. The difference at the end of three minutes in EMG signals from the biceps, deltoideus, and trapezius muscles when: (A) the hand is firmly grasping the middle of the telephone receiver, holding it against the ear, while the upper arm and elbow are kept away from the chest at an angle of about 60°; and (B) the hand is holding the lower end of the telephone receiver, and the elbow is resting against the chest.

Integrated electromyographic recordings permit the measurement of the load on single muscles, as well as groups of muscles. This can be done equally as well both by surface electrodes and by intramuscular electrodes. It is well established that the EMG reflects the magnitude of the muscle engagement, and may be used to measure the exerted forces in per cent of the maximal voluntary muscle strength (Hagberg, 1981; Figure 14.2).

When a working muscle becomes fatigued, characteristic changes occur in the EMG. These changes include a change in impulse traffic in the innervating motor nerve, and in the reaction of the muscle fibers to the electrical discharge. With the onset of fatigue the amplitude increases and the rhythm slows down (Komi, 1983).

Muscle tension in static industrial work

Neuromuscular complaints of the kind previously described, and which are generally considered to be the result of prolonged, intense static work, are by all indications extremely common. Thus, in a questionnaire survey, Tola *et al.* (1988) found that about half of the 1,194 machine operators questioned had had neck and shoulder symptoms during the preceding seven days, as against only 24 per cent of the office workers.

Westgaard *et al.* (1984) made a survey of muscular-skeletal complaints and EMG-assessed muscular load in young seamstresses in four different factories. Although they were doing similar work, the absence

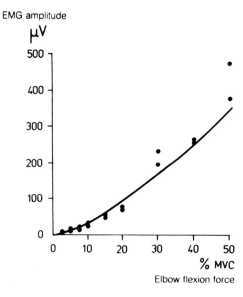

Figure 14.2. Relationship between EMG-activity from the biceps brachii muscle and elbow flexion force as a percentage of the maximal voluntary contraction (% MVC) (from Hagberg, 1981).

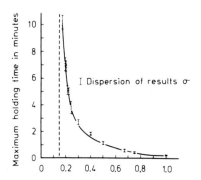

Figure 14.3. The maximal time a force in isometric muscular contraction can be maintained, expressed in percentage of maximal isometric strength (from Rohmert, 1968).

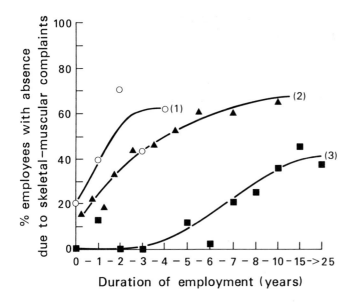

Figure 14.4. Percentage employees with absence due to skeletal-muscular complaints, in relation to duration of employment: (1) workers at an electronic assembly plant, (2) seamstresses, (3) office workers (modified from Westgaard et al., 1984).

due to muscular-skeletal complaints was quite different in the different factories. Skeletal-muscular or neuromuscular complaints accounted for between one-quarter and one-half of the total sick absence in these four factories. The sick absence due to skeletal-muscular complaints was not related to age. Within three years of employment, close to half of the seamstresses had been absent from work for prolonged periods because of skeletal-muscular complaints (Figure 14.4). The load on the trapezius muscle, assessed by EMG recordings, was measured in some of the subjects at two of the clothing factories by using a stationary electromyograph. The median static work load was about 6 per cent MVC, varying between 3 and 9 per cent MVC (Westgaard *et al.*, 1984).

In spite of major efforts and considerable investment in order to improve the working conditions ergonomically in the factories, no convincing evidence of permanently reduced muscular work load or reduced absence due to skeletal-muscular complaints was observed. In two of the factories the absence due to neuromuscular complaints increased, in one it remained the same, and in one it dropped (Westgaard *et al.*, 1984).

The fact that it is not only the working posture that plays a role in causing neuromuscular tension leading to skeletal-muscular complaints is indicated by the findings of Tola *et al.* (1988). They showed that besides working in twisted or bent postures, age and job satisfaction also proved to be significant risk indicators for neck and shoulder symptoms. It should also be kept in mind that body posture in general may affect the relative load on different muscle groups. With the centre of gravity of the head and trunk very close to the supporting column of bones, Man has the most economical anti-gravity mechanism among the mammals, once the upright posture is attained. During forward flexion of the spinal column there is a marked muscular activity until flexion is extreme, at which point ligament structures take over the support of the trunk, and the electromyographically recorded discharge from the trunk muscles ceases (Floyd and Silver, 1955; Basmajian, 1979).

Mental tension, and myoelectrical feedback

At rest, the skeletal muscles are relaxed, and normally no electrical activity can be recorded by electromyography. Mental tension, however, even in the inactive, resting individual, will cause a recordable EMG activity, more or less proportional to the degree of tension (Waersted et al., 1987, 1988), similar to the increase in heart rate (Figure 13.15) and blood pressure (Figure 13.19).

Chronic tension in certain muscle groups may not only cause pain but also cramps, such as writer's cramps, or temporo-mandibular jaw syndrome, stiff neck, aching shoulders and tension headache. It should also be kept in mind that prolonged static contraction not only has an untoward effect on the individual muscle fibers. It also affects the blood supply to the entire musculature involved due to the lack of venous pumping which naturally takes place during dynamic muscular work.

Not everyone is equally able to sense the state of tension of different muscle groups in their own body. In order to assist individuals in developing such muscular sensitivity, a variety of myoelectrical feedback instruments have been developed.

The function of myoelectric feedback instruments is to pick up and amplify the myoelectrical voltage oscillations which appear over the muscle as it changes its state of tension, and to present them as signals, which the users can employ to improve control of their muscles. By setting a tension threshold combined with an audible signal such an instrument can be used to let the subject know when the tension of the muscle being monitored exceeds a certain limit, and to tell the person to relax.

Several versions of such bio-feedback instruments are available. One is the pocket-sized M402, produced by Aleph One Ltd., Cambridge, England, using three skin electrodes to pick up signals from the surface of the body. It has calibrated controls which determine the desired level of a signal required to obtain feedback. This can take the form of a beep feedback, proportional to the tension above threshold, or an alarm feedback when tension exceeds the threshold. This can be combined with a logger such as the Squirrel (Grant, Cambridge, England) for the storing, treatment, display and plotting of the data. This combination with a Squirrel provides an ambulatory system which can be used for the purpose of logging the tension in certain muscle groups over extended periods, even 24 hours, to find out when and why excessive tension develops in the muscle groups being monitored. For further details, see Basmajian (1983, 1985), and datas sheets 2, 9, 12 and 14 from Aleph One Ltd., Cambridge, England.

Conclusion

From this brief summary it is evident that the kind of continuous static work which is common in certain types of industrial operations is harmful to a large number of operators, even at loads less than 10 per cent MVF. Evidently, the rate of work output, or the level of productivity in such operations, plays a role in the development of muscular tension leading to neuromuscular complaints, causing prolonged sick absence from work. It is true that a number of operators do not develop such complaints, but it takes several months of exposure to find out whether or not an individual is prone to develop such tension-related complaints. It is therefore difficult to perform a preventive selection of suitable operators who are unlikely to develop such complaints.

Practically speaking, therefore, the ideal solution of this problem would obviously be to relieve human operators of this kind of static work, leaving it to mechanical devices to perform operations of this kind. In the cases where this is not possible, the work routine should be altered, so as to allow more varied use of muscle groups, and thereby reduce the length of each period of static contractions. Ideally, static work of the kind discussed here should be alternated with varied dynamic work at reasonable intervals. In the establishment of such varied work routines, including reasonable intervals between static and dynamic work, the ambulatory EMG recording system now available may be a useful tool in arriving at objective criteria for the establishment of individual work routines. Such ambulatory monitoring would also be useful in finding out the causative relationships between muscle tension and the development of neuromuscular complaints, and why some individuals do, and others do not, develop such complaints when engaged in the same occupation.

Chapter 15
Working in a carbon monoxide polluted atmosphere

Introduction

Carbon monoxide (CO) is produced in any kind of incomplete combustion. This happened around the primitive fireplace of the stone age hunter in his cave. It happens today in the campfire of the modern hunter in the woods, in many industrial operations, and in a burning cigarette.

The noxious effect of the carbon monoxide is mainly due to its choking effect on vital cells which, in extreme cases, may be lethal to the organism. The real problem is that carbon monoxide has an affinity to hemoglobin, which is more than 200 times greater than that of oxygen. Since it is the hemoglobin which carries the oxygen to the different cells and tissues of the body, the amount of CO taken up replaces an equivalent amount of oxygen. Thus CO is taken up at the expense of oxygen, hence its choking effect. And since the blood is so keen on picking up CO, it also hangs on to it that much longer. This means that habitual smokers have a tendency to accumulate greater amounts of CO as carbon monoxyhemoglobin (COHb) at the expense of a corresponding amount of oxygen, reducing endurance correspondingly. This is the reason why top athletes in endurance events do not smoke. The smokers who inhale, are especially apt to accumulate COHb, but even passive smokers are affected.

The amount of CO taken up and distributed within the body depends on several factors:

1. It depends on the CO-content in the ambient air, which ideally is negligible but which in extreme cases may reach concentrations of several hundred ppm.
2. The CO-uptake also depends on the length of time which the individual is exposed to the ambient air CO-concentration in question (see Figure 15.1).

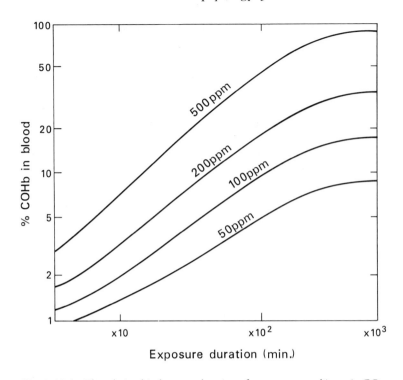

Exposure duration (min.)

Figure 15.1. The relationship between duration of exposure to ambient air CO concentration in ppm and COHb content of the blood expressed as per cent saturation (modified from Peterson and Stewart, 1970).

3. Finally, the CO-uptake depends on the amount of CO-containing air which is brought down into the subject's lungs (pulmonary ventilation), and thus brought in contact with his blood, to be attached to the hemoglobin molecule. At rest, the pulmonary ventilation of an average individual may be less than 10 liters per min; at maximal effort it may be in excess of 100 liters per min. (For references, see Åstrand and Rodahl, 1986.)

There is a linear relationship between work rate and pulmonary ventilation: the heavier the work rate, the greater the pulmonary ventilation. Similarly, the heart rate increases linearly with work rate. Therefore heart rate may be used, indirectly, as an index of pulmonary ventilation.

The combined effect of the above mentioned factors (CO-content of the inhaled air, duration of exposure and pulmonary ventilation) will determine how much CO is accumulated in the blood as COHb. And this can be determined by a simple biochemical analysis of a small blood sample drawn from the subject. In the final analysis, this is the parameter that matters.

The purpose of this chapter is to discuss the possibility of assessing each one of the above mentioned factors, and determining how the COHb content of the blood may be assessed without having to take any blood samples. This will include the application of a practical method of recording and logging CO-exposure, and CO-uptake in workers in CO-exposed industry. Basically, this approach is based on the use of sensors for the measurement of pertinent environmental parameters combined with the recording of physiological reactions of the exposed individual to these environmental parameters. Furthermore, it is based on the continuous storing of all these parameters in the internal memory of the Squirrel logger (Grant, Cambridge, England) carried on the person. In this manner the pertinent environmental parameters are measured where the subject is located at any time, and the results are immediately available.

Assessment of the CO-content of the ambient air

Three different CO-sensors were used for the purpose of measuring the carbon monoxide content of the ambient air at a typical silicon carbide (SiC) production plant in Norway, the Norton Co. at Lillesand. They were the British Sabre sensor (output -20 to $+20$ mV), and the two German sensors: the Compur (output -200 to $+200$ mV) and the Dräger (output -2 to $+2$ V).

All three sensors were calibrated against known test gases containing 50 and 121 ppm CO in air. An example of the calibration curve is presented in Figure 15.2, which shows a straight line relationship between the concentration of CO in ppm in the test gas and the mV reading on the Squirrel, which was quite similar in all three sensors. A re-calibration several weeks later showed almost identical results, indicating a high degree of stability of all three sensors.

Detailed records of CO-concentrations at different locations in the Norton SiC-producing plant, both by permanently installed recording instruments and by ambulatory CO-meters (Dräger Commonwarn), have been kept over a number of years. From these records it appears that the average concentrations of CO in the passage between the two rows of ovens may vary roughly between 20 and 80 ppm, and somewhat lower and more stable levels around 30–40 ppm in the adjacent hall where the finished SiC is sorted out.

At the time of our field test, a general survey of the CO-concentrations in the ambient air inside the plant was made with the aid of the conventional Dräger Commonwarn portable battery-operated non-recording CO-meter. This instrument, which sucks samples of the

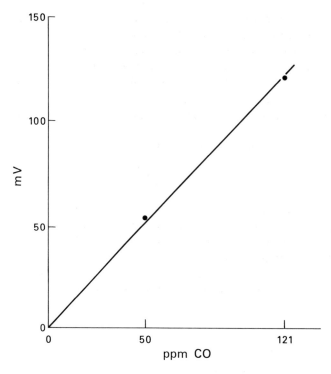

Figure 15.2. A straight-line relationship between the concentration of CO in the test gases and the mV reading on the Squirrel meter/logger using a Compur CO-sensor.

ambient air into the reacting chamber with the aid of a pump, in contrast to the three test sensors, which are supplied with a passive membrane for electrochemical CO-sensing, is extremely fast reacting, but does not record the readings, nor does it store them in an electronic memory. This survey showed CO-concentrations between 20 and 45 ppm in the sorting hall. Similar readings, between 30 and 40 ppm, were observed in a sheltered area in the middle of the passage between the two rows of SiC-ovens in the production hall, while readings between 55 and 90 ppm were observed in the actual passage between the ovens. Here extremely rapid and very marked fluctuations in the CO-content of the ambient air were observed. At times the CO-concentration might fluctuate as much as 50 ppm in a matter of seconds. This makes any kind of recording by a mechanical indicating meter difficult, a problem which may be augmented by human factors, including the problem of deciding what meter readings to record. This is another reason why a portable, ambulatory logger system may be preferable, especially when the logger is set to record every second or every few seconds.

Following this initial survey, the CO-concentration in the ambient air was measured, one meter above the floor at a number of specific locations in the production hall and in the sorting hall, where the finished metal is sorted according to specifications, every five seconds by the Dräger Commonwarn, compared with the Squirrel-connected sensors logging every second. As long as the ambient air was fairly stable, as was the case in the sorting hall, the agreement between the four sensors was fairly good, as is evident from the following mean results:

Dräger Commonwarn: 24 ppm; Sabre: 28 ppm
Dräger Commonwarn: 38 ppm; Compur: 36 ppm
Dräger Commonwarn: 50 ppm; Dräger: 56 ppm

There were considerable differences, however, when the air was turbulent, as in the passage between the ovens in the production hall, due to draft or moving hot objects in the vicinity. As already mentioned, the difference was most probably due to the fact that the Dräger Commonwarn reacted much faster because of the pump. In the passage between the ovens in the production hall, the recordings gave the following mean results:

Dräger Commonwarn: 87 ppm; Sabre: 123 ppm
Dräger Commonwarn: 79 ppm; Compur: 79 ppm
Dräger Commonwarn: 83 ppm; Dräger: 104 ppm

It is thus evident that at the lower, and fairly stable CO-concentrations, all three Squirrel-attached test-sensors were usable for the practical purpose of assessing ambient CO-concentrations. The main difficulty, as already pointed out, were the very rapid and marked fluctuations in the CO-content of the ambient air, especially in the production hall (see Figure 15.3).

In an attempt to overcome this problem, and to obtain stable samples for comparative measurements, samples of ambient air were collected at two typical locations in the plant. The air was collected in large airtight plastic bags, with the aid of a pump. This trapped air was then thoroughly mixed and analyzed by the standard Dräger Commonwarn meter and the Ecolyser meter in the laboratory. These readings were then compared with the readings of the three Squirrel-attached test sensors. To secure an even flow, a constant weight was placed on top of the bag. The results were as follows:

Bag no. 1: CO-content measured by:
Dräger Commonwarn: 130 ppm
Sabre sensor: 145 ppm
Compur sensor: 130 ppm
Dräger sensor: 136 ppm

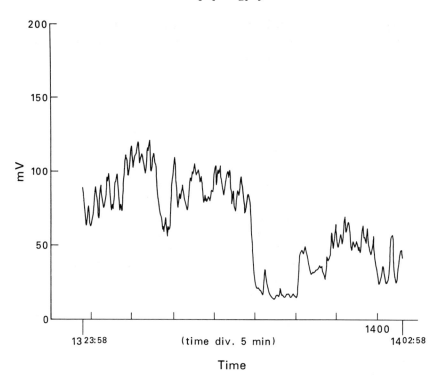

Figure 15.3. CO-concentration in the ambient air in a SiC production plant, in mV, recorded by the Compur sensor plugged into the Squirrel meter/logger (1 mV is roughly equal to 1 ppm CO).

Bag no. 2: CO-content measured by:
Dräger Commonwarn: 38 ppm
Ecolyser: 40 ppm
Sabre: 47 ppm
Compur: 43 ppm
Dräger: 46 ppm

The conclusion to be drawn from this field test is that under specified conditions (low air movement, and moderate CO-concentrations), any one of these three sensors may be used to record and to log the CO-content of the ambient air.

Assessment of pulmonary ventilation

The pulmonary ventilation expressed in liters per min can be measured directly by the Douglas bag method, or indirectly by recording the heart rate, as an indication of work load or work rate which affect pulmonary ventilation. If fairly accurate values are required, the above mentioned correlations may be established in each of the subjects in separate experiments, prior to the actual field study at the individual work place. For further details see Åstrand and Rodahl, 1986, Chapter 5.

Assessment of actual CO-uptake (by the determination of the carbon monoxyhaemoglobin (COHb) content of the blood)

The percentage COHb saturation of the blood can, as mentioned, be assessed directly by the analyses of venous blood samples by conventional methods. However, it is well established that alveolar breath analysis may be used for a fairly accurate estimation of the per cent carboxyhaemoglobin (COHb) saturation in healthy individuals, provided the ambient CO-concentration is near zero (Peterson, 1970; Rea et al., 1973; Rees et al., 1980; Jarvis et al., 1980; Wald et al., 1981). Coburn, Forster and Kane (1965) developed a theoretical equation for the prediction of changes in COHb per cent saturation levels, which is quite accurate for constant levels of environmental CO and constant pulmonary ventilation (Peterson and Stewart, 1975; Hauck and Neuberger, 1984; Muller and Barton, 1987). This situation, however, is rarely encountered in real life, and certainly not in CO-exposed industry (Tikuisis et al., 1987).

The Bedfont Carbon Monoxide Monitor, Model EC 50, is an inexpensive, easy to operate instrument, specifically designed to measure CO-concentrations in the expired air, as an indication of COHb content, expressed as per cent saturation. Jarvis et al. (1986), have compared the results obtained by this monitor with measured blood carboxyhaemoglobin concentrations in 72 subjects whose COHb-concentrations ranged from 0.3 to 12 per cent. The expired air CO-concentrations measured by the monitor correlated closely with the COHb-concentrations, the relationship being linear over this range. As an approximate guide, the percentage of COHb-saturation could be obtained by dividing the expired air reading by six. In their experience, the second reading on the Bedfont EC 50 monitor gave somewhat higher values than the first reading. Irving et al. (1988) have compared the same instrument with the Ecolyzer in 138 normal subjects and confirmed the findings of Jarvis et al. (1986).

In the above mentioned studies the emphasis was on the application of the Bedfont monitor as an instrument to estimate tobacco smoking status, and as a motivational device for those attempting to stop smoking. Our interest in the monitor stems from the need to provide occupational health personnel in carbon monoxide-exposed industry with a reliable instrument for assessing COHb-saturation in their workers following exposure on their jobs. For this reason, the Bedfont CO monitor was connected with the 1201 Squirrel meter/logger, supplied with an event marker for the logging of individual readings at specific times.

In a pilot study carried out at the Pulmonary Department of Haukeland Hospital in Bergen, in collaboration with Dr P. Bakke and Dr T. Guthe, the instrument was used to record CO-concentration in ppm in the expired air from two subjects who were non-smokers, one subject who said that she only smoked 10–15 cigarettes per day, and one subject who said that he smoked in excess of 25 cigarettes per day. The study was performed in a laboratory where the CO-content of the ambient air was negligible. One subject was tested at a time, seated in a chair. Prior to the actual test, which was repeated twice, the subject was instructed in the use of the monitor and allowed to practice. A venous blood sample was drawn from the cubital vein for COHb-analysis by a *OSM3 Hemoximeter* (Zijlstra et al., 1988). The subject then exercised on a cycle ergometer at 100 Watt for 5 minutes, a blood sample was drawn, and the expired air was measured by the CO-monitor. The subject then exercised again at 150 or 200 Watt for 5 minutes, following which the same testing procedure was repeated.

The procedure for the expiratory air CO-measurements was as follows: the subject was asked to exhale, then to inhale as deeply as possible, holding the breath for 20 seconds (recorded by stopwatch), and then to exhale into the Bedfont mouthpiece forcefully and as completely as possible, pressing the elbows against the chest. The readings on the Bedfont monitor (ppm) were continuously recorded, and compared with the mV-readings on the Squirrel.

The results showed a straight line relationship between ppm in the expired air and the photometrically measured percentage COHb saturation, and between the Squirrel readings in mV and the measured percentage COHb saturation (Figure 15.4). From Table 15.1, it appears that the difference between the estimated COHb saturation, based on the expired air CO, according to the graph supplied with the instrument, and the actually measured percentage COHb saturation in the blood sample, varies from 0.3 to 0.7 per cent. On the average it appears that the Bedfont CO-monitor in our hands underestimated the carboxyhemoglobin saturation by 0 6 per cent. Roughly speaking, it appears that the percentage COHb saturation may be obtained by dividing the expiratory air CO-concentration in ppm by about 5 (4.7 to be exact), or by dividing the Squirrel reading in mV by 7 (actually 6.9).

This pilot study was followed by a field study at a typical CO-exposed work place (the Norton SiC company), in which the Bedfont CO-monitor was tested in combination with sensors for the assessment of CO-content of the ambient air, and a heart rate recorder to determine work rate, plugged into the same Squirrel meter/logger. The

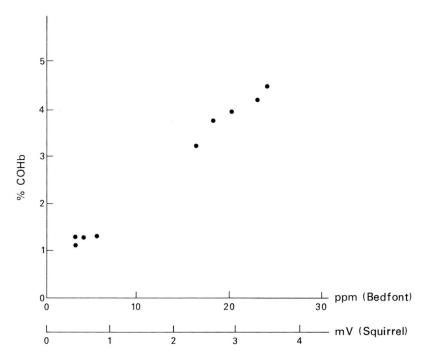

Figure 15.4. A straight line relationship between CO content in expired air (read in ppm on the Belfont CO-monitor, Model EC 50, and in mV on the Squirrel meter/logger) and the photometrically measured percentage COHb saturation.

Table 15.1. CO (ppm) in alveolar air, estimated and measured COHb% saturation in two non-smokers and two smokers at rest and following exercise.

Subject No.	Activity	CO in exp. air ppm	COHb% Estimated	COHb% Measured	COHb% Difference
1	rest	6	1.0	1.3	0.3
2	rest	4	0.7	1.3	0.6
	100 W	3	0.6	1.3	0.7
	200 W	3	0.6	1.2	0.6
3	rest	20	3.3	4.0	0.7
	100 W	18	3.0	3.8	0.6
	150 W	16	2.7	3.2	0.5
4	rest	24	4.0	4.5	0.5
	100 W	23	3.8	4.3	0.5
	150 W	20	3.4	4.0	0.6

results of the Bedfont-predicted percentage COHb saturation were compared with the results of the analysis of COHb per cent saturation in blood samples from the same subjects taken at the same time.

The subjects participating in the field study were exposed to different work loads for 15 minutes using a cycle ergometer to ensure constant and repeatable work loads. In view of the fact that heavy tobacco smokers may have as much as 5–10 per cent COHb saturation due to the CO-uptake from the tobacco smoke alone, both smokers and non-smokers were used as subjects for this field study. The six subjects who volunteered for the study were all employees of the company, working in the CO-atmosphere of the production plant. Three of them were smokers, and three were non-smokers. They reported at 9 a.m. at the laboratory where the CO-content of the ambient air was close to zero, without having been exposed to the CO-containing atmosphere of the SiC-production area earlier that day. They remained in the CO-free laboratory area until their turn came to take part in the test exposure. Prior to the CO-exposure, blood samples were drawn for the analysis of the percentage COHb saturation at the Pulmonary Department at Haukeland Hospital in Bergen (Dr P. Bakke). The CO-content of the alveolar air was then measured by the Bedfont CO-monitor as previously described, for the prediction of per cent COHb saturation. The Compur CO-sensor was attached to the subject's coat, close to his neck, connected to the Squirrel carried by the subject with the aid of a belt. The Squirrel was set to record the CO-content once a minute, to be compared with stationary recording of the CO-content of the ambient air by another Compur-sensor attached to a separate Squirrel logger, placed about 1.5 meters away from the cycle ergometer where the subject was to be situated during the test. A heart rate recorder connected to the Squirrel carried by the subject was mounted on the subject, for the continuous logging of his work rate. The subject than walked out to the SiC-production plant, where subjects 1–5 spent 15 min each in the sorting hall, while subject 6 went to the adjacent end of the production hall, where the CO-content of the ambient air was considerably higher. Subjects 1–3 were smokers, subjects 4–6 were non-smokers. The physical activity of the individual subjects is given in Table 15.2, where the mean heart rate during the 15-min period is also listed. The CO-content of the ambient air was continuously recorded and logged once every second by the stationary sensor, and read once every 10 seconds on a Dräger CO-meter placed next to the Compur sensor located at the level of the subject's head about 1.5 meters away.

The two Compur CO-sensors used were calibrated prior to the test (50 mV = 50 ppm CO). The heart rate recorder had been checked against ECG recordings and manual pulse counts.

At the end of the 15-min exposure period the subject returned to the laboratory where a blood sample was drawn for COHb saturation analysis, following which the alveolar CO–content was assessed by the Bedfont CO–monitor as previously described. In this case the test was repeated three times, and the highest values were used. As a supplementary test, one smoker and one non-smoker were tested by the Bedfont analyzer before and after a 10–minute walk along the passage between the two rows of SiC ovens.

Table 15.2. Work rate and heart rate during a 15-minute test.

Subject No.	Work rate	Heart rate Squirrel-logged
1	rest	99
2	300 kpm/min	96
3	450 kpm/min	103
4	600 kpm/min	127
5	600 kpm/min	97
6	600 kpm/min	96

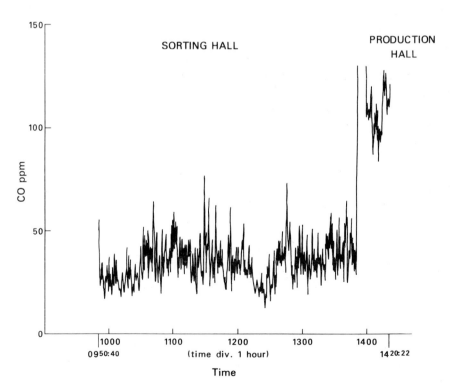

Figure 15.5. CO-concentration in the ambient air (ppm) in the sorting hall and the production hall, logged once every second (1 mV = 1 ppm).

CO-exposure

The results of the stationary recording of the CO-content of the ambient air once every second during the entire period of the tests are presented in Figure 15.5. Two extreme examples of a more detailed presentation of the CO-exposure values for subjects 1 and 6 for their 15-min exposure periods are given in Figures 15.6 and 15.7. The mean values and the range for the CO-concentrations are given in Table 15.3 (subjects 1–5 in the sorting hall, and subject 6 in the production hall).

As is evident from Table 15.3 and the Figures 15.5–15.7, there are great and extremely rapid fluctuations in the CO-content of the ambient air surrounding the subject. It is also evident that the CO-content is markedly lower in the sorting hall than in the production hall. It appears from Table 15.3 that the values for CO-content recorded and logged by the Compur sensor every second, and the values read by the Dräger meter once every 10 seconds, are very similar. In contrast to this there are considerable differences between the Compur sensor values recorded once every second, and the values

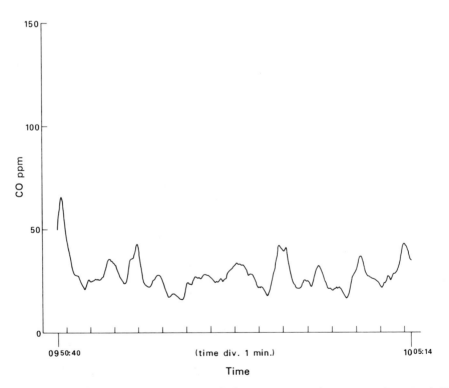

Figure 15.6. Showing CO-exposure (ppm) of subject no. 1, resting for 15 min in the sorting hall, Norton SiC Co.

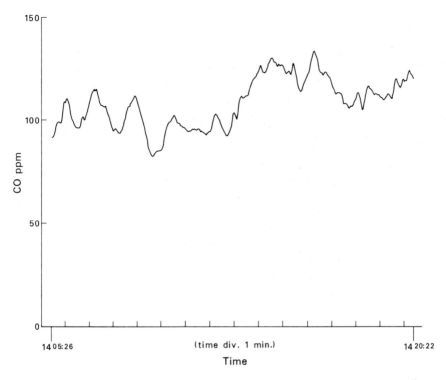

Figure 15.7. Showing CO-exposure (ppm) of subject no. 6 exercising at 600 kpm · min⁻¹ for 15 min in the production hall, Norton SiC Co. (CO-logging once every second).

Table 15.3. CO-concentration (ppm) in the ambient air during exposure. Mean values, min. and max. values in parentheses.

Subject no.	Recorded with Compur II 1 recording/sec	Measured with Dräger 1 reading/10 sec	Recorded with Compur I 1 recording/min
1	28 (16–66)	25 (12–43)	19 (12–31)
2	36 (18–68)	35 (13–79)	41 (17–92)
3	36 (21–54)	34 (18–60)	26 (19–33)
4	33 (16–66)	31 (16–78)	20 (–30)
5	23 (10–47)	23 (11–42)	19 (14–30)
6	107 (80–133)	94 (71–119)	70 (60–80)

recorded once every minute from the Compur sensor carried on the subject. The main reason for this discrepancy is probably that measurements once a minute are not frequent enough under the existing conditions of rapid fluctuations of the CO-content. However, the possibility that the levels and fluctuations of the CO-concentrations were different at the two sites, i.e. at the location of the subject and the location of the two stationary sensors 1.5 m away, cannot be excluded.

There are large individual differences in the recorded heart rate at the same work rate (Table 15.2). This is obviously due to large individual differences in the level of physical fitness.

COHb saturation before and after CO-exposure in smokers and non-smokers, at different work rates

Table 15.4 shows the values for the percentage COHb saturation in the blood, predicted on the basis of measured CO-content in the alveolar air, compared with the results of biochemical COHb analyses, before and after CO-exposure in the six subjects. It appears that the mean difference between predicted and actually measured COHb saturation in the blood is no more than one per cent before and 0.7 per cent after the CO-exposure. The difference is quite small at the lower values, i.e. in the non-smokers, subjects 1–3. It is conceivable that part of the difference in smokers prior to the CO-exposure may be caused by the blood samples being drawn some time before the Bedfont test, and that the COHb saturation might have declined somewhat during the time period between blood sample and Bedfont test, assuming that the subjects had smoked just prior to arriving at the laboratory.

These results, however, support the findings of the pilot study which showed that, for practical purposes, the Bedfont method is quite applicable for the purpose of predicting the CO-saturation level in the blood of workers in the CO-exposed industry.

In all the three subjects who were tobacco smokers, the percentage

Table 15.4. Per cent COHb saturation

Subject no.	Before exposure		After exposure	
	Predicted	Measured	Predicted	Measured
1	1.2	2.8	1.9	2.5
2	6.0	8.5	5.8	7.3
3	2.4	3.2	2.5	2.8
4	0.9	1.1	1.4	1.5
5	1.0	1.5	1.4	1.5
6	0.9	1.0	2.2	2.2

COHb saturation was higher before than after CO-exposure. This may perhaps, again, be partly due to the subjects having smoked shortly before arriving at the laboratory, thus having a high COHb level to start with, which then declined during the subsequent period of time which elapsed. This drop in COHb saturation, which most probably did occur, would probably offset the increase in COHb saturation during the experimental CO-exposure.

One of the non-smokers had the same COHb saturation before and after the CO-exposure, the remaining two non-smokers had higher COHb levels after exposure. The subject who had the greatest increase in COHb saturation was subject 6 who was exposed to considerably higher CO-concentration than the rest of the subjects (80–133 ppm as against 10–66 ppm).

The difference in the basic pre-exposure levels of COHb saturation between smokers and non-smokers is much greater than the difference in COHb saturation before and after the CO-exposure. One of the smokers had a COHb saturation of 8.5 per cent before the CO-exposure, while the greatest increase in COHb saturation observed during the CO-exposure was only 1.2 per cent. This was in subject 6 who was exposed to a mean CO-concentration in the ambient air of 107 ppm for 15 min and a work load of 600 kpm per min. It is thus evident that tobacco smoking may be more important than brief periods of industrial exposure to carbon monoxide in terms of affecting the CO-accumulation in the blood.

The significance of the subject's work rate during CO-exposure under the conditions of this experiment is not quite clear. Nor do we know if the work loads used in this experiment are representative of the work loads to which the workers are exposed during their ordinary daily work in the production and sorting halls. Subject 4 who had a higher work pulse than subject 5 at identical work loads, also had a greater increase in the percentage COHb saturation, but subject 4 also had a higher CO-exposure than did subject 5 (33 as against 23 ppm).

The results of the COHb saturation tests of the two individuals who spent 10 min walking along the pathway between the two rows of ovens in the production hall showed an increase in the predicted percentage COHb saturation from 1.0 to 1.8 in the non-smoker, and from 4.2 to 4.7 in the smoker. This again emphasizes the role played by habitual tobacco smoking in accumulating COHb in the blood. It also indicates that even brief periods of exposure to the prevailing levels of CO-concentrations in the ambient air in some industrial plants may cause a noticeable COHb accumulation.

Conclusion

The observations discussed here, as well as previous reports, show a reasonable degree of agreement between the values for COHb saturation in individual cases, predicted on the basis of the Bedfont CO-Monitor, and the values obtained by the analysis of the COHb saturation in individual blood samples. It may thus be concluded that the Bedfont technique may be used to assess the percentage COHb saturation in the blood in CO-exposed individuals with a satisfactory degree of accuracy. The technique is simple and can be used by the industry's own technicians without scientific training, after a brief course of instruction. The interpretation of the results is also quite simple. Furthermore, it is possible to make a similar assessment of COHb saturation as that achieved by the Bedfont monitor by using any Squirrel-compatible CO-sensor such as those mentioned above (Sabre, Compur, Dräger, etc.), supplied with a proper mouthpiece and gas chamber placed over the sensor membrane connected to a Squirrel meter/logger. In this manner, the same CO-sensor may be used both to log the CO-content of the ambient air, and to assess the percentage COHb saturation in the exposed subject's blood. Furthermore, it is possible with available technology to produce a direct reading COHb per cent meter, based on alveolar CO, logged on an electronic logger, such as the Squirrel, sufficiently accurate for monitoring both smoking status and for monitoring COHb levels in workers exposed to CO. Ideally, such a device should provide readouts of both CO-levels in the atmosphere at the work place, and the predicted COHb per cent saturation, based on alveolar CO.

These observations also clearly illustrate the advantage of using an IBM compatible logger with an electronic memory. This allows an immediate display of the results, and the opportunity of data analysis and printout. This represents a major advantage compared to a meter or a monitor which only permits the reading and manual recording of the values as they are displayed.

By this logger system it is also possible to measure and log the physical work rate of the CO-exposed individual. This permits a systematic analysis of the relative importance of this parameter. It is also technically possible to supplement the instrument assembly with sensors or probes which measure the pulmonary ventilation directly, for instance by using flow meters compatible with the Squirrel meter/logger.

By far the most striking observation made so far in this project is the important role played by the individual's smoking habits for the maintenance of a high COHb accumulation in the blood, both at work and outside his place of work. The fact that tobacco smoking alone may cause higher levels of COHb saturation than brief periods of industrial CO-exposure is noteworthy. A more systematic and detailed investigation of this matter would be of considerable interest, particularly from an occupational health point of view.

The fact that a 10-min walk between the rows of SiC-ovens may cause a small but noticeable increase in the percentage COHb saturation suggests that it might be worthwhile to survey the CO-concentrations in the ambient air more systematically by this technique in order to provide a more objective and realistic basis for a further evaluation of different operational routines.

The main purpose of this project was to contribute to the development and field testing of a method for an objective survey of the CO-environment, combined with a non-invasive method of estimating the percentage COHb saturation of the blood, as a meaningful expression of the physiological consequences of the CO-exposure. While this objective, in essence, has been attained, the project has also brought up a number of practical and theoretical questions which may be answered by the use of the applied method. This includes a more systematic and detailed mapping of the CO-exposure and the effect this may have on the COHb saturation of the workers at the different work places. It would be equally interesting to find out how long it takes for the increased COHb saturation to return to its initial level, following the exposure. It would also be of interest to examine further the effect of the work rate on the CO-uptake, and to what extent the level of physical fitness may influence this relationship.

References

Aanonsen, A., 1959, Medical problems of shift-work. *Ind. med. surg.*, **28**, 422.

Aanonsen, A., 1964, *Shift work and health* (Oslo: Universitetsforlaget).

Abrams, C.F. jr., and Suggs, C.W., 1977, Development of a simulator for use in the measurement of chain saw vibration. *Appl. Ergonomics*, **8**(3), 130–4.

ACGIH (American Conference of Governmental Industrial Hygienists), *Threshold limit values for chemical substances and physical agents in the workroom environment with intended changes for 1976* (Cincinnati, Ohio).

Adolph, E.F., 1947, *Physiology of man in the desert.* Monographs in the Physiological Sciences (New York: Interscience Publishers, Inc.).

Åkerstedt, T., 1985, Shifted sleep hours. *Ann. Clin. Res.*, **17**(5), 273.

Åkerstedt, T. and Torsvall, L., 1981, Shift work, shift-dependent well-being and individual differences. *Ergonomics*, **24**(4), 265.

Alm, N.O. and Rodahl, K., 1979, "Work stress of Norwegian coal miners in Spitsbergen" (Unpublished MS).

Althouse, R. and Murrell, J.J., 1977, *An analysis of job stress in coal mining* (NIOSH Research Report, U.S. Dept. of Health, Education and Welfare, Cincinnati, Ohio).

Alvarez, W.C. and Stanley, L.L., 1930, Blood pressure in six thousand prisoners and four hundred prison guards. *Arch. Int. Med.*, **46**, 17.

Andauer, P. and Metz, B., 1967, Le travail en équipes alterantes, in J. Scherrer (ed.), *Physiologie du travail-Ergonomie*, (Paris: Masson et Cie.), p. 272.

Anderson, R.M. jr. and Bremer, D.A., 1987, Sleep duration at home and sleepiness on the job in rotating twelve-hour shift workers. *Human Factors*, **29**(4), 477.

Andersson, B., Houmøller, S., Karlson, C-G., and Svensson, S., 1974, Analysis of urinary catecholamines: an improved auto-analyzer fluorescence method. *Clin. Chem. Acta*, **51**, 13–28.

Andersson, B., Istance, H. and Spencer, J. (eds), 1977, *Human factors in the design and operation of ships*. Paper presented to the First Internat. Conf. on Human Factors in the Design and Operation of Ships, Gothenburg.

Andersson, L., 1983, *Intervention against loneliness. Effects on blood pressure*. Sixth Nordic Congress of Gerontology, Copenhagen, Denmark, 10–13 April.

Angersbach, D., Knauth, P., Loskant, H., Karvonen, M.J., Undeutsch, K. and Rutenfranz, J., 1980, A retrospective cohort study comparing complaints and diseases in day and shift workers. *Int. Arch. Occup. Environ. Health*, **34**(2), 127.

Armstrong, T.J., Fine, L.J., Radwin, R.G. and Silverstein, B.S., 1987, Ergonomics and the effects of vibration in hand-intensive work. *Scand. J. Work Environ. Health*, **13**(4), 286.

Asken, M.J. and Raham, D.C., 1983, Resident performance and sleep deprivation: a review. *J. Med. Education*, **58**(5), 382–8.

Åstrand, I., 1960, Aerobic work capacity in men and women with special reference to age. *Acta Physiol. Scand.*, **49**(Suppl. 169).

Åstrand, I., Guharay, A. and Wahren, J., 1968, Circulatory responses to arm exercise with different arm positions. *J. Appl. Physiol.*, **25**, 258.

Åstrand, I., Axelson, O., Eriksson, U. and Olander, L., 1975, Heat stress in occupational work. *AMBIO*, **4**(1), 37–42.

Åstrand, I., Åstrand, P.-O., Christensen, E.H. and Hedman, R., 1960, Intermittent muscular work. *Acta Physiol. Scand.*, **48**, 443.

Åstrand, I., Åstrand, P.-O., Hallback, I. and Kilbom, A., 1973a, Reduction in maximal oxygen uptake with age. *J. Appl. Physiol.*, **35**, 649.

Åstrand, I., Fugelli, P., Karlsson, C.G., Rodahl, K. and Vokac, Z., 1973b, Energy output and work stress in coastal fishing. *Scand. J. Clin. Lab. Invest.*, **31**, 105.

Åstrand, P.-O., 1952, Experimental studies of physical working capacity in relation to sex and age (Copenhagen: Ejnar Munksgaard).

Åstrand, P-O., 1982, How to conduct and evaluate an exercise stress test, in I. Kass (ed.) *International Approaches to Issues in Pulmonary Disease*, Monograph 18, p. 20 (New York: World Rehabilitation Fund Inc.).

Åstrand, P.-O. and Christensen, E.H., 1964, Aerobic work capacity, in F. Dickens, E. Neil and W.F. Widdas (eds), *Oxygen in the Animal Organism* (New York: Pergamon Press), 295.

Åstrand, P.-O. and Rodahl, K., 1986, *Textbook of Work Physiology, 3rd edition* (New York: McGraw Hill).

Åstrand, P-O. and Ryhming, I., 1954, a normogram for calculation of aerobic capacity (physical fitness) from pulse rate during submaximal work. *J. Appl. Physiol.*, **7**, 218.

Åstrand, P.-O., Ekblom, B., Messin, R., Saltin, B. and Stenberg, J., 1965, Intra-arterial blood pressure during exercise with different muscle groups, *J. Appl. Physiol.*, **20**, 253.

Baird, D.D., 1985, Occupational exposure to heat or noise and reduced fertility. *JAMA*, **253**(18), 1643.

Baker, D., 1980, The use and health consequences of shift work. *Int. J. Health Services*, **10**(3), 405.

Baker, M.A., 1982, Brain cooling in endotherms in heat and exercise. *Ann. Rev. Physiol.*, **44**, 85.

Banister, E.W. and Brown, S.W., 1968, The relative energy requirements of physical activity, in *Exercise Physiology*, (New York: Academic Press Inc.).

Barnes, B.L., 1983, Personality profile of merchant marine officers. *Indian J. Clin. Psychol.*, **10**, 423.

Basmajian, J.V., 1979, *Muscles Alive* (Baltimore: Williams & Wilkins).

Basmajian, J.V., 1983, *Biofeedback: principles and practice for clinicians* (Baltimore: Williams & Wilkins).

Basmajian, J.V., 1985, *Muscles Alive: Their Function Revealed by Electromyography* (Baltimore: Williams & Wilkins).

Behnke, A.R., 1978, Physiological effects of abnormal atmospheric pressures, in G.D. Clayton and F.A. Clayton (eds), *Patty's Industrial Hygiene and Toxicology, 3. rev. ed., vol. 1, General Principles,* (New York: John Wiley and Sons Inc.).

Benzinger, T.H. and Taylor, G.W., 1963, Cranial measurements of internal temperature in man, in J.D. Hardy (ed.), *Temperature: Its Measurement and Control in Science and Industry, vol. 3, Part 3* (New York: Reinhold Book Corp.), p. 111.

Bergh, U., 1980, Human power at subnormal body temperatures. *Acta Physiol. Scand.* (Suppl. 478).

Beshir, M.Y., Ramsey, J.D. and Burford, C.L., 1982, Threshold values for the Botsball: a field study of occupational heat. *Ergonomics,* **25**(3), 247.

Bigland, B. and Lippold, O.C.J., 1954, The relation between force, velocity and integrated electrical activity in human muscles. *J. Physiol.,* **123**, 214.

Bjørklund, R.A. and Heien, N. 1987, "Manual for Selection of Maritime Crew, Version 1.0" (Unpublished).

Bjørklund, R.A., Rodahl, K., Robertson, B.J., Wøhni, B. Otherhals, O., Øverås, P., Vespestad, J. and Midtgard, Å., 1987, *Effects of Maritime Simulator Training on Performance.* Proc. 4th Internat. Conf. on Marine Simulation. The Internat. Marine Simulator Forum, Trondheim, Norway, June 22–24.

Björksten, M. and Jonsson, B., 1977, Endurance limit of force in long-term intermittent static contractions. *Scand. J. Work Environ. & Health,* **3**, 23.

Black, D.W. 1984, Laughter, *J.A.M.A.,* **252**(21), 2995.

Blix, I., Bolling, A. and Rodahl, K., 1979, *En arbeidsfysiologisk undersøkelse om bord på M/S* Vindafjord *på fart fra Afrika til Europa,* 3 S-rapportserien, Oslo.

Bobet, J. and Norman, R.W., 1982, Use of the average electromyogram in design evaluation. Investigation of a whole-body task, *Ergonomics,* **25**(12), 1155.

Bobo, N., Bethea, N.J., Ayoub, M.M. and Intaranont, K., 1983, Energy expenditure and aerobic fitness of Malew low seam coal miners. *Human Factors,* **25**(1), 43.

Bondarev, G.I., Zinovien, E. Sh., Nepoklonov, Y.U.A. and Endovitskaya, I.S., 1962, Energy expended by fish-processing men engaged on board fish trawlers cruising in the waters of the Barents Sea and North Atlantic. *Vop. Pitan.,* **21**, 40 (in Russian).

Booze, C.F., 1979, Morbidity experience of air traffic control personnel – 1967–77. *Aviation Space & Environ. Med.,* **50**, 1–8.

Botsford, J.H., 1971, A wet globe thermometer for environmental heat measurement. *AIHA J.,* **32**, 1.

Brief, A.P., 1980, How to manage managerial stress, *Personnel,* **57**(5), 25.

Brief, R.S. and Confer, R.G., 1971, Comparison of heat stress indices, *Am. Ind. Hyg. Assoc. J.,* **32**, 11.

Brinnel, H., Nagasaka, T. and Cabanac, M., 1987, Enhanced brain protection during passive hyperthermia in humans, *Eur. J. Appl. Physiol,* **56**, 540.

Brotherhood, J.R., 1973, Studies on energy expenditure in the Antarctic, in O.G. Edholm and E.K.E. Gunderson (eds), *Polar Human Biology,* (London: Heinemann Medical Books Ltd.), 182.

Brotherhood, J.R., 1985, Snow, cold and energy expenditure: a basis for fatigue and skiing accidents. *The Australian J. Science and Medicine in Sports*, **17**(1), 3.

Brotherhood, J.R., 1987, The practical assessment of heat stress in J.R.S. Hales and D.A.B. Richards (eds), *Heat Stress, Physical Exertion and Environment*. Proc. 1. World Conf. Heat Stress, Sydney, Australia (Amsterdam: Elsevier Science Publ., Excerpta Medica), 451.

Brown, G.A. and Williams, G.M., 1982, The effect of head cooling on deep body temperature and thermal comfort in man. *Aviation, Space, and Environmental Medicine*, 583.

Browne, R.C., 1949, The day and night performance of teleprinter switchboard operators. *Occup. Psychol.*, **23**, 121.

Bugge, J.F., Opstad, P.K. and Magnus, P.M., 1979, Changes in the circadian rhythm of performance and mood in healthy young men exposed to prolonged, heavy physical work, sleep deprivation, and caloric deficit. *Aviat. Space Environ. Med.*, **50**(7), 663–8.

Burton, A.C., 1935, Human calorimetry II – the average temperature of the tissues of the body, *J. Nutr.* **9**, 261.

Burton, A.C., 1963, The pattern of response to cold in animals and the evolution of homeothermy, in J.D. Hardy (ed.), *Temperature: Its Measurement and Control in Science and Industry, Vol. 3, Part 3* (New York: Reinhold Book Corp.), 363.

Burton, A.C. and Edholm, O.G., 1955, *Man in a Cold Environment* (London: Edward Arnold).

Burton, A.C. and Edholm, O.G., 1969, *Man in a Cold Environment*, (New York and London: Hafner Publishing Co.).

Cabanac, M., 1986, Keeping a cool head. *News in Physiological Sciences*, **1**, 41.

Cabanac, M., Germain, M. and Brinnel, H., 1987, Tympanic temperatures during hemiface cooling, *Eur. J. Appl. Physiol.*, **56**, 534.

Cadenhead, R. McN., 1976, Hospital admissions of fishermen from the fishing grounds around the Shetland Islands. *J. Soc. Occup. Med.*, **26**, 127.

Cannon, W.B., 1929, *Bodily Changes in Pain, Hunger, Fear and Rage*, (New York: Appleton & Co).

Carruthers, M., Arguelles, A.E. and Mosovich, A., 1976, Man in transit: biochemical and physiological changes during intercontinental flights. *The Lancet*, May 8, 977.

Chakraborty, M.K., Sensarma, S.K. and Sarkar, D.N., 1974, Blood lactic acid in determining the heaviness of different mining work. *Indian J. Physiol. Pharmacol.*, Oct–Dec., 341.

Christensen, E.H., 1960, Muscular work and fatigue, in K. Rodahl and S.M. Horvath (eds), *Muscle as a Tissue* (New York: McGraw-Hill Book Co.).

Christensen, E.H. and Högberg, P., 1950, Physiology of skiing. *Arbeitsphysiologie*, **14**, 292.

Ciriello, V.M. and Snook, S.H., 1977, The prediction of WBGT from the Botsball. *Am. Ind. Hyg. Assoc. J.*, **38**, 264.

Cobb, S. and Rose, R.M., 1973, Hypertension, peptic ulcer, and diabetes in air traffic controllers, *J.A.M.A.*, **224**, 489.

Coburn, R.F., Forster, R.E. and Kane, P.B., 1965, Considerations of the physiological variables that determine the blood carboxyhaemoglobin concentration in man. *J. Clin. Invest.*, **44**(11), 1899.

Collacott, R.A., 1977, Risks to trawler fishermen in Orkney waters. *J. Royal Coll. Gen. Pract.*, **27**(181), 482.

Collins, K.J,. Eddy, T.P., Hibbs, A., Stock, A.L. and Wheeler, E.F., 1971, Nutritional and environmental studies on an ocean-going oil tanker. 2. Heat Acclimatization and Nutrient Balances. *Brit. J. Industr. Med.*, **28**, 246.

Colquhoun, W.P., 1970, Circadian rhythms, mental efficiency and shift work, *Ergonomics*, **13**(5), 558.

Colquhoun, W.P., 1971, Circadian variations in mental efficiency, in *Biological Rhythms and Human Performance* (London: Academic Press, Inc.).

Colquhoun, W.P., Blake, M.J.F. and Edwards, R.S., 1968a, Experimental studies of shift-work I: a comparison of a 'Rotating' and 'Stabilized' 4-hour shift systems. *Ergonomics*, **11**(5), 437.

Colquhoun, W.P., Blake, M.J.F. and Edwards, R.S., 1968b, Experimental studies of shift-work II: stabilized 8-hour shift systems. *Ergonomics*, **11**(6), 527.

Colquhoun, W.P., Blake, M.J.F. and Edwards, R.S., 1969, Experimental studies of shift-work III: stabilized 12-hour shift systems. *Ergonomics*, **12**(6), 865.

Comroe, J.H. jr., 1966, The Lung, *Sci. Am.*, **214**, 56.

Consolazio, C.F., 1963, The energy requirement of men living under extreme environmental conditions. *World Rev. of Nutrition and Dietetics*, **4**, 53.

Consolazio, C.F., Johnson, R.E. and Pecora, L.J., 1963, *Physiological Measurements of Metabolic Functions in Man* (New York: McGraw-Hill Inc).

Cook, T.C. and Cashman, P.M.M., 1982, Stress and ectopic beats in ships' pilots. *J. Psychosomatic Res.*, **26**(6), 559.

Cooper, C.L. and Melhuish, A., 1980, Occupational stress and managers. *J. Occup. Med.*, **22**(9), 588.

Court, A., 1948, Wind chill. *Bull. Am. Met. Sc.*, **29**, 487.

Courtice, F.C. and Douglas, C.G. 1936, "The Effects of Prolonged Muscular Exercise on the Metabolism". *Proceedings of the Royal Soc. London*, Vol. 119, 381–439.

Cox, S., Cox, T., Thirlaway, M. and Mackay, C., 1982, Effects of simulated repetitive work on urinary catecholamine excretion. *Ergonomics*, **25**(12), 1129.

Crane, R.A., 1978, Digital socially acceptable monitoring instrument, *Med. & Biol. Eng. & Comput.*, **16**, 78.

Cross, T. 1985, The health of British trawlermen on the Arctic fishing grounds. *J. Soc. Occup. Med.*, **35**, 55.

Crowdy, J.P., 1970, Hunger patrol in M.S. Peterson (ed.), *Feeding the Military Man* (U.S. Army Natic Labs, Mass.).

Crump, J.H., 1979, Review of stress in air traffic control, its measurement and effects. *Aviat. Space Environ. Med.*, **50**(3), 243.

Czeisler, C.A., Weitzman, E.D., Moore-Ede, M.C., Zimmerman, J.C. and Knauer, R.S., 1980, Human sleep: its duration and organization depend on its circadian phase. *Science*, **210**, 1264.

Dalgaard, J.B., Dencker, F., Fallentin, B., Hansen, P., Kaempe, B., Steensberg, J. and Wilhardt, P., 1970, *Sundhetsstyrelsen*, **5**(17), 209.

Dalgaard, J.B., Dencker, F., Fallentin, B., Hansen, P., Kaempe, B., Steensberg, J. and Wilhardt, P., 1972, Fatal poisoning and other health hazards connected with industrial fishing. *Brit. J. Industr. Med.*, **29**(3), 307.

Darwin, C., 1890, *Expression of the Emotions in Man and Animals, 2nd ed.* (New York: Appleton & Co.).

Davenport, H.W., 1981, Signs of anxiety, rage or distress. *The Physiologist,* **24**(5), 1.

David, M. and Popescu, M.P., 1972, Changes in ocular diopters under conditions of work at high temperatures. *Fiziologia Normala si Patologica, Romanio,* **18**(5), 439.

Day, R., 1949, Regional heat loss, in L.W. Newburgh (ed.) *I. Physiology of Heat Regulation* (Philadelphia: W.B. Saunders Co.), 240.

Diamond, J., 1983, The biology of the wheel, *Nature,* **302** (5909), 572.

Dibeneditto, J.P. and Worobec, S.M., Exposure to hot environments can cause dermatological problems. *Occup. Health and Safety,* **54**(1), 35.

Dinman, B.D. and Horvath, S.M., 1984, Heat disorders in industry *J. Occup. Med.,* **26**, 489.

Dobronravova, N.P., 1962, Energy expenditure of fishermen – natives of Arctic regions. *Vop. Pitan,* **21**, 36 (in Russian).

Dolmierski, R. and Walden, de K.J., 1972, Clinical evaluation of suicides by seamen. *Psychiatria Polska,* **6**(3), 255.

Dolmierski, R., Nitka, J. and Mrozinski, W., 1986, Evaluation of remote effects of ship's bombardment on the state of health of the crew members, *Bull. Inst. Marit. Trop. Med. Gdynia,* **37**, 25.

Dougherty, J.D., 1967, Cardiovascular findings in air traffic controllers. *Aerospace Med.* **38**, 26–30.

Dougherty, J.D., Trites, D.K. and Dille, J.R., 1965, Self-reported stress-related symptoms among air traffic control specialists (ATCS) and non-ATCS personnel. *Aerospace Med.* **36**, 96–960.

Dukes-Dobos, F.N., 1981, Hazards of heat exposure. *Scand. J. Work Environ. Health,* **7**, 73.

Durnin, J.V.G.A. and Passmore, R., 1967, *Energy, Work and Leisure* (London: William Heinemann, Ltd.).

Eaton, S.B. and Konner, M., 1985, Paleolithic nutrition, *New England J. Med.,* **312**(5), 283.

Ebert, H., 1979, Health protection on board. Training programme for ship's officers. *Bull. Inst. Marit. Trop. Med. Gdynia,* **30**(1), 9.

Egeland, T., Lossius, P., Enger, N., Gundersen, N., Bolling, A. and Jebens, E., 1981, *Varmebelastningsundersøkelse ved Fiskaa Verk. Kristiansand.* Arbeidsfysiologisk Inst., Oslo.

Ekblom, B. and Huot, R., 1972, Response to submaximal and maximal exercise at different levels of carboxyhemoglobin, *Acta Physiol. Scand.,* **86**, 474.

Ekenvall, L., 1987, Clinical assessment of suspected damage from handheld vibrating tools, *Scand. J. Work Environ. Health,* **13**(4), 271.

Ellis, F.P., 1970, Medical aspects of trawler safety, *Brit. J. Industr. Med.,* **27**(1), 78.

Elo, A.-L., 1979, *Seamen's Work and Mental Health – A Literature Review.* Turku Regional Inst. Occup. Health, Helsinki, Finland.

Engel, P., Henze, W., Armonies, G. and Munscher, M., 1984, Psychological and physiological performance during long-lasting work in heat with and without wearing cooling vests, in *Thermal Physiology*, ed. J.R.S. Hales (New York: Raven Press).

Erikssen, J., Rasmussen, K., Forfang, K. and Storstein, O., 1977, Exercise ECG and case history in the diagnosis of latent coronary heart disease among presumably healthy middle-aged men. *Eur. J. Cardiol.*, **5–6**, 463.

Erikssen, J., Forfang, K. and Jervell, J., 1981a, Coronary risk factors and physical fitness in healthy middle-aged men. *Acta. Med. Scand.* (Suppl. 645), 57.

Erikssen, J., Johansen, A.H. and Rodahl, K., 1981b, Coronary heart disease in Norwegian sea-pilots; part of the occupational hazard? *Acta Medica Scand.* (Suppl. 645).

Erikssen, J., Knudsen, K.E., Guthe, T., Lützow-Holm, J.P. and Rodahl, K., 1988, Blood pressure of heat-exposed workers at Fiskaa Verk. Unpublished results.

Euler, U.S. von, 1964, Quantification of stress by catecholamine analysis. *Clin. Pharm. and Therapeutics*, **5**, 398–404.

Euler, U.S. von, and Hellner, S., 1952, Excretion of noradrenaline and adrenaline in muscular work, *Acta Physiol. Scand.* **26**, 183–91.

Euler, U.S. von and Lishajko, F., 1961, Improved technique for the fluorimetric estimation of catecholamines, *Acta Physiol. Scand.*, **51**, 348–56.

Fagraeus, L., 1974, Cardiorespiratory and metabolic functions during exercise in hyperbaric environment. *Acta Physiol. Scand.* (Suppl. 414).

Fernandez, R.H.P., 1980, Health care of people at work, *J. Soc. Occup. Med.* **30**, 40.

Filikowski, J., 1977, Health condition of seamen of ocean-going transport vessels. *Bull. Inst. Marit. Trop. Med. Gdynia*, **28**, 107.

Filikowski, J., 1978, Effect of the working environment on ocean-going ships, on the seamen's health. Suggestions concerning prophylaxis indications. *Bull. Inst. Marit. Trop. Med. Gdynia*, **29**, 5.

Filikowski, J., 1979, Effect of obesity on the morbidity of seamen. *Bull. Inst. Marit. Trop. Med. Gdynia*, **30**(1), 5.

Finberg, J.P.M. and Berlyne, G.M., 1977, Modification of renin and aldosterone response to heat by acclimatization in man. *J. Appl. Physiol. Respirat. Environ. Exercise Physiol.*, **42**(4), 554–8.

Flook, V., 1987, Physics and physiology in the hyperbaric environment. *Clin. Phys. Physiol. Meas.*, **8**(3), 197.

Floyd, W.F. and Silver, P.H.S., 1955, The function of the erectores spinae muscles in certain movements and postures of man. *J. Physiol.*, **129**, 184.

Folk, G.E. jr., 1974, *Textbook of Environmental Physiology, 2nd ed.* (Philadelphia: Lea and Febiger).

Folkard, S. and Monk, T.H. (eds), 1985, *Hours of Work. Temporal Factors in Work-Scheduling*, in *Psychology and Productivity at Work Series* (New York: John Wiley & Sons).

Folkard, S., Monk, T.H. and Lobban, M.C., 1979, Towards a predictive test of adjustment to shift work. *Ergonomics*, **22**(1), 79.

Folkard, S., Minors, D.S. and Waterhourse, J.M., 1985, Chronobiology and shift work: current issues and trends. *Chronobiologia*, **12**(1), 31.

Fowler, F.D., 1980, Air traffic control problems: a pilot's view. *Human Factors*, **22**(6) 645–53.

Francis, K.T., 1979, Psychologic correlates of serum indicators of stress in man: a longitudinal study. *Psychosom. Med.* **41**(8), 617.

Frankenheuser, M., 1968, *Physiological and Subjective Reactions to Different Physical Work Loads*. Report from the Psychological Laboratories, University of Stockholm, No. 254.

Fröberg, J., Karlsson, C.G., Levi, L. and Lidberg, L., 1972, Circadian variations in performance, psychological ratings, catecholamine excretion, and diuresis during prolonged sleep deprivation, *Intern. J. Psychobiol.*, **2**, 23–36.

Fugelli, P.I., 1979a, Vaerøy og Røst – prosjektet II. *Tidsskrift for Den Norske Laegeforening*, **99**(28), 1394.

Fugelli, P.I., 1979b, Vaerøy og Røst – prosjektet III. *Tidsskrift for Den Norske Laegeforening*, **99**(29), 1463.

Funderburk, C.F., Hipskind, S.S., Welton, R.C. and Lind, A.R., 1974, Development of and recovery from fatigue induced by static effort at various tensions. *J. Appl. Physiol.*, **37**(3), 392.

Gaddie, J., Legge, J.S., Friend, J.A.R. and Reid, T.M.S., 1980, Pulmonary hypersensitivity in prawn workers. *Lancet*, **2**(8208), 1350.

Gagge, A.P., Burton, A.C. and Bazett, H.C., 1941, A practical system of units for description of heat exchange of man with his environment, *Science*, **94**, 428.

Gagge, A.P., Gonzalez, R.R. and Nishi, Y., 1974, Physical factors governing man's thermal comfort, discomfort and heat tolerance, *Build. Internat.* **7**, 305–31.

Gemne, G., 1987, Symptomatology and diagnostic methods in the hand-arm vibration syndrome. *Scand. J. Work Environ. Health*, **13**(4), 271–388.

Gerathewohl, S.J., 1973, The role of man in system safety. *Review de Medicine Aeronautique et Spatiale, No. 49.*

Gisolfi, C.V. and Copping, J.R., 1974, Thermal effects of prolonged treadmill exercise in the heat. *Med Sci. Sports*, **6**, 108.

Goethe, H., Zorn, E., Herrmann, R. and Bernd-Fred Schepers, 1978a, Psychophysical load of personnel on modern sea-going ships as a present problem in nautical medicine. *Z. bl. Bakt. Hyg. I. Abt. Orig. B.* **166**, 1–36.

Goethe, H., Schmidt, E.G., Zorn, E., Rinas, W., Herrmann, R. and Schmidt, H., 1978b, Lärm und Schiffen. *Schiff und Hafen/Kommandobrücke*, **4**, 321.

Goldman, R.F., 1974, Clothing design for comfort and work performance in extreme thermal environments. *New York Acad. Sci. Trans.*, **36**(6), 531.

Grandjean, E., 1961, Musik und Arbeit. *Zeitschrift für Präventivmedizin*, **6**, 65–70.

Grandjean, E.P., Wotzka, G., Schaad, R. and Gilgen, A., 1971, Fatigue and stress in air traffic controllers. *Ergonomics*, **14**(1), 159.

Greenleaf, J.E. and Castle, B.I. 1971, Exercise temperature regulation in man during hypohydration and hyperhydration. *J. Appl. Physiol.*, **30**, 847.

Greenleaf, J.E., Van Beaumont, W., Brock, P.J., Montgomery, L.D., Morse, J.T., Shvartz, E. and Kravik, S., 1980, Fluid-electrolyte shifts and thermoregulation: Rest and work in heat with head cooling. *Aviation, Space and Environmental Medicine*, August, p. 747.

Grimby, G. and Saltin, B., 1983, The ageing muscle. *Clin. Physiol.*, **3**, 209.

Grout, J.W., 1980, Occupational stress of intensive care nurses and air traffic controllers: review of related studies. *J. Nursing Educ.*, **19**, No. 6, 8.

Guidry, A.J. and McDowell, R.E., 1966, Tympanic membrane temperature for indicating rapid changes in body temperature. *J. Dairy Science*, **49**(1), 74.

Gusic, B., Krajina, Z., Pljak, Z., Konic-Carnelutti, V. and Babic, I., 1969, The influence of heat on the upper respiratory tracts. *Acta Otolaryngologica*, **67**, 150.

Hagberg. M., 1981, On Evaluation of Local Muscular Load and Fatigue by Electromyography. Arbete och Hälsa Vetenskaplig Skriftserie Arbetarskyddsverket, Stockholm, Sweden.

Hale, H.B., Williams, E.W. and Smith, B.N., 1971, Excretion patterns of air traffic controllers. *Aerospace Med.* **41**, 127–38.

Halsey, M.J., 1982, Effects of high pressure on the central nervous system, *Physiol. Rev.*, **62**(4), 1341.

Hampton, T.R.W., 1974, Selection, detection and rejection. The assessment of fitness for stress. *J. Roy. Nav. Med. Serv.*, **60**, 88–92.

Hanson, J.A. and Jones, F.P., 1970, Heart rate and small postural changes in man, *Ergonomics*, **13**(4), 483.

Hardy, J.D., 1967, Central and peripheral factors in physiological temperature regulation, in *Les Concepts de Claude Bernhard sur le Milieu Intërieur* (Paris: Masson et Cie), 247.

Harris, C.S., Sommer, H.C. and Johnson, D.L., 1976, Review of the effects of infrasound on man. *Aviat. Space Environ. Med.* **47**(4), 430–4.

Harrison, M.H., Brown, G.A. and Belyavin, A.J., 1982, The 'Oxylog': an evaluation. *Ergonomics*, **25**(9), 809.

Hauck, H. and Neuberger, M., 1984, Carbon monoxide uptake and the resulting carboxyhemoglobin in man. *Eur. J. Appl. Physiol.*, **53**, 186.

Hausman, A. and Petit, J.M., 1979, Physical work in a hot environment in coal mines. *Annales des mines de Belgique*, **11**, 1109.

Hauty, G.T. and Adams, T., 1966a, Phase shifts of the human circadian system and performance deficit during the periods of transition: 1. East-West Flight. *Aerospace Med.*, **37**, 668.

Hauty, G.T. and Adams, T., 1966b, Phase shifts of the human circadian system and performance deficit during the periods of transition: II. West-East Flight. *Aerospace Med.*, **37**(10), 1027.

Hauty, G.T. and Adams, T., 1966c, Phase shifts of the human circadian system and performance deficit during the periods of transition: III North-South Flight. *Aerospace Med.*, **37**(12), 1257.

Hawkins, L.H. and Armstrong-Esther, G.A., 1978, Circadian rhythms and night shift working in nurses. *Nursing Times*, **74**(18), 49.

Helgason, T., Asmundsson, G., Broddason, T., Olafsson, H., Hannesdottir, H. and Stefansson, J.G., 1977, Havfiskere og Deres Familier. *Tidsskrift for Den Norske Laegeforening*, **97**(27), 1389.

Herd, J.A., 1972, The physiology of strong emotions: Cannon's scientific legacy re-examined. Sixteenth Bowditch Lecture. *The Physiologist*, **15**(1), 5.

Hernberg, S., Karava, R., Koskela, R. and Luoma, K., 1976, Angina pectoris, ECG findings and blood pressure of foundry workers in relation to carbon monoxide exposure. *Scand. J. Work Envir. & Health*, **2**, suppl. 1, 54–63.

Hettinger, T. and Rodahl, K., 1960a, A work physiology study of an assembly line operation. *J. Occup. Med.*, **2**(11), 532–5.

Hettinger, T. and Rodahl, K., 1960b, Ein Modifizierter Stufentest Zur Messung Der Belastungsfähigkeit Des Kreislaufes. *Deutsche Medizinische Wochenschrift*, **14**, 553.

Hieber, A.R. and Smidt, U., 1982, Ergometrische Untersuchungen an Bergleuten und Hüttenarbeitern zur Differenzierung der Ursachen einer Belastungsdyspnoe. *Praxis Pneumalogie*, **36**.

Hildebrandt, G., Rohmert, W. and Rutenfranz, J., 1974, 12 & 24 H Rhythms in Error Frequency of Locomotive Drivers and the Influence of Tiredness. *Internat. J. Chronobiol.*, **2**(2), 175.

Hilfert, von R., Köhne, G., Toussaint, R. and Zerlett, G., 1980, Die Kontinuierliche mobile Erfassung und digitale Speicherung von Biosignalen in Arbeitsmedizin und Ergonomi, *Zentralblatt für Arbeitsmedizin Arbeitsschutz und Prophylaxe*, **30**(3), 57.

Hill, A.V., 1939, *Living Machinery* (London: G. Bell & Sons).

Hinkle, L.E. jr., 1973, The concept of 'stress' in the biological and social sciences. *Science, Medicine and Man*, **1**(1), 31.

Hitchen, M., Brodie, D.A. and Harness, J.B., Cardiac responses to demanding mental load. *Ergonomics*, **23**(4), 379.

Hoiberg, A. and McCaughey, B.G., 1984, The traumatic after effects of collision at sea. *Am. J. Psych.* **141**, 70.

Holmboe, J., Bell, H. and Norman, N., 1975, Urinary excretion of catecholamines and steroids in military cadets exposed to prolonged stress, *Försvarsmedicin*, **11**, 183.

Holmér, I., 1981, *Role of Clothing for Man's Heat Exchange with the Environment*. Proc. Intern. Conf. on Protective Clothing Systems, Stockholm, Sweden, Aug. 23–27, 31–50.

Holmér, I. and Elnäs, S., 1983, Physiological evaluation of the resistance to evaporative heat transfer by clothing. *Ergonomics*, **24**(1), 63.

Hong, S.K., Rennie, D.W. and Park, Y.S., 1987, Humans can acclimatize to cold: a lesson from Korean women divers. *New in Physiological Sciences*, **2**, 79.

Hoodless, D.J., Macdonald, I.A., Ballal, M.A. and Blecher, A., 1980, A portable physiological data recording/decoding system, *J. Biomed. Engng*, **2**, 193.

Hopkins, C.O., How much should you pay for that box? *Human Factors*, **17**(6), 533.

Höyem-Johansen, A. and Natvig, N.L., 1978, *Arbeidsbelastningen og Stresshormon-utskillelsen hos norske loser*, 3 S-rapport serien, Oslo.

I.L.O., ATC/1979/10 Report, Meeting of Experts on Problems Concerning Air Traffic Controllers, Geneva, 8–16 May.

Irving, J.M., Clark, E.C., Crombie, I.K. and Smith, W.C.S., Evaluation of a portable measure of expired-air carbon monoxide. *Prev. Med.*, **17**, 109.

ISO (International Organization for Standardization), 1971, *ISO Recommendation R 1999 Acoustics. Assessment of Occupational Noise Exposure for Hearing Conservation Purposes*.

ISO 1974, *Guide for the Evaluation of Human Exposure to Whole-Body Vibration*. ISO 2631. Geneva.

ISO 1979, *Guide for the Measurement and the Evaluation of Human Exposure to Vibration Transmitted to the Hand*. ISO 5349. Geneva.

Issekutz, B. jr., Birkhead, N.C. and Rodahl, K., 1962, Use of respiratory quotients in assessment of aerobic work capacity. *J. Appl. Physiol.*, **17**, 47.

Iwata, H. (ed.), 1974, Body reaction by vibration, in *Vibration Syndrome* (Tokyo: Kindaishuppann Ltd.), 21.

Jansen, T., Olsen, B., Hatlevold, M., Enger, N., Guthe, T., Johannessen, H. and Rodahl, K., 1982, *En undersøkelse av varmeeksponeringen ved Lista Aluminiumverk.* AFYI. Publikasjon, Arbeidsfysiologisk Inst., Oslo, Norway.

Jaremin, B., Chmielewski, J., Bielawska-Krasnowiecka, G., Nahorski, W., Glombiowski, P. and Winnicka, A., 1981, The effect of gastric and/or duodenal ulcers and the complications arising from them on the course of the professional work of seamen and ocean-going fishermen. *Bull. Inst. Marit. Trop. Med. Gdynia*, **32**, 169.

Jarvis, M.J., Russell, M.A.H. and Saloojee, Y., 1980, Expired air carbon monoxide: a simple breath test of tobacco smoke intake. *Brit. Med. J.*, **281**, 484.

Jarvis, M.J., Belcher, B., Vesey, C. and Hutchison, D.C.S., 1986, Low cost carbon monoxide monitors in smoking assessment. *Thorax*, **41**, 886.

Jebava, R., Hrochova, J. and Rencova, E., 1980, Glassblower's cataract. *Czekoslovenska Oftalmologie*, **36**(2), 115.

Jebens, E. and Bolling, A., 1980, *En arbeidsfysiologisk undersøkelse av varmebelastningen hos maskinister og reparatører om bord på M/S Tarago i fart på den Persiske Gulf.* 3 S-rapport serien, Oslo.

Jelliffe, D.B., 1966, *The Assessment of the Nutritional Status of the Community.* World Health Organization, Geneva, 238.

Jokl, E., 1980, Professor A.V. Hill. A personal tribute. *J. Sports Med.*, **20**(4).

Jonsen, J. (ed.), 1976, Sjøfartsmedisinsk forskning. Nordisk Utredningsserie 1976: 11 Rapport fra nordisk symposium, Oslo, 16–17 juni 1975. Utgiven av Nordiska rådet och Nordiska ministerrådets sekretariat i Oslo, Stockholm.

Jonsson, A. and Hansson, L., 1977, Prolonged exposure to a stressful stimulus (noise) as a cause of raised blood pressure in man. *The Lancet*, January 8, 86.

Jungman, H., 1960, Kreislauf und Körpertemperatur während einer Schiffsreise in die Tropen. *Arch. Phys. Ther.*, **12** (no. 3), 225–33.

Kamon, E., 1974, Instrumentation for work physiology. *Trans. N.Y. Acad. Sci.*, no. 7, **36**, 625.

Kamon, E., 1979, Scheduling cycles of work for hot ambient conditions. *Ergonomics*, **22**(4), 427.

Kamon, E. and Ryan, C., 1981, Effective heat strain index using pocket computer. *Am. Ind. Hyg. Assoc. J.*, **42**(8), 611.

Karson, S. and O'Dell, J.W., 1974, Personality makeup of the American air traffic controller. *Aerospace Med.*, **45**(9), 1001–07.

Kaufman, J.E. and Christensen, J.F. (eds.), 1972, *IES Lighting Handbook, 5th ed.* (New York: Illuminating Engineering Society).

Kerslake, D. McK., 1963, Errors arising from the use of mean heat exchange coefficients in calculation of the heat exchange of a cylindrical body in a transverse wind, in J.D. Hardy (ed.), *Temperature: Its Measurement and Control in Science and Industry*, Vol. 3, part 3 (New York: Reinhold Book Corp.), 183.

Kerslake, D.McK., 1969, The value of reflecting layers in clothing. *Proc. R. Soc. Med.* (England), **62**(3), 283.

Kerslake, D. Mck., 1972, The stress of hot environments. *Monogr. Physiol. Soc.*, **29** (Cambridge University Press).

Kersten, E. von, 1967, Ergebnisse Arbeitshygienisker Analysen Berufsspezifischer Gesundheitsgefahren in der Modernen Hochseefischerei. *Sante Publique*, **8**(4), 361.

Khogali, M. and Hales, J.R.S., (eds.), 1983, *Heat Stroke and Temperature Regulation* (New York: Academic Press).

Kilbom, Å., 1969, *Lotsar och Lotsningsarbete*, Arbetsmed. Inst. Stockholm.

Kjellberg, A. and Wikström, B.-O., 1985, Whole-body vibration: exposure time and acute effects – a review. *Ergonomics*, **28**(3), 535.

Klein, K.E., Brüner, H., Holtmann, H., Rehme, H., Stolze, J., Steinhoff, W.D. and Wegmann, H.M., 1970, Circadian rhythm of pilots' efficiency and effects of multiple time zone travel. *Aerospace Med.*, **41**(2), 125.

Kleitman, N., 1963, *Sleep and Wakefulness* (Chicago: University of Chicago Press).

Kloetzel, K., Etelvino de Andrade., A., Falleiros, J. and Cota Pacheco, J., 1973, Relationship between hypertension and prolonged exposure to heat. *J. Occup. Med.*, **15**, No. 11, 878–80.

Knauth, P. and Rutenfranz, J., 1976, Experimental shift work studies of permanent night, and rapidly rotating, shift systems. 1. Circadian rhythm of body temperature and re-entrainment at shift change. *Int. Arch. Occup. Environ. Health*, **37**(2), 125.

Knauth, P., Rutenfranz, J., Hermann, G. and Poeppl, S.T., 1978, Re-entrainment of body temperature in experimental shift-work studies. *Ergonomics*, **21**(10), 775.

Knauth, P., Landau, K., Dröge, C., Schwitteck, M., Dynskl, M.W. and Rutenfranz, J., 1980a, Duration of sleep depending on the type of shift work. *Int. Arch. Occup. Environ. Health*, **46**(2), 167.

Knauth, P., Rutenfranz, J., Schulz, H., Bruder, S., Romberg, H.P., Decoster, F. and Kiesswetter, E., 1980b, Experimental shift work studies of permanent night, and rapidly rotating shift systems. II. Behaviour of various characteristics of sleep. *Int. Arch. Occup. Environ. Health*, **46**(2), 111.

Knave, B. and Floderus, B., 1988, Exposure to low-frequency electromagnetic fields – a health hazard? *Scand. J. Work Environ. Health*, **14**(suppl. 1), 46.

Knochel, J.P., Beisel, W.R., Herndon, E.G., Gerard, E.S. and Barry, K.G., 1961, The renal, cardiovascular, hematologic and serum electrolyte abnormalities of heat stroke, *Amer. J. Med.*, **299**.

Knutsson, A., Åkerstedt, T. and Jonsson, B.G., 1988, Prevalence of risk factors for coronary artery disease among day and shift workers. *Scand. J. Work Environ. Health*, **14**, 317.

Kolmodin-Hedman, B. and Swensson, Å., 1975, Problems related to shift work. A field study of Swedish railroad workers with irregular work hours. *Scan. J. Work. Environ. & Health*, **1**(4), 254.

Komi, P.V., 1983, Electromyographic, mechanical, and metabolic changes during static and dynamic fatigue, in *Biochemistry of Exercise*, Int. Series of Sport Sciences, Vol. 13 (Champaign, IL: Human Kinetics Publishers Inc.), 197.

Konz, S. and Gupta, V.K., 1969, Water cooled hood affects creative productivity, *Ashrae J.*, July, 40.

Kosunen, K.J., Pakarinen, A.J., Kuoppasalmi, K. and Adlercreutz, H., 1976, Plasma renin activity, angiotension II and aldosterone during intense heat stress. *J. Appl. Physiol.*, **41**(3), 323–7.

Kotchen, T.A., Hartley, L.H., Rice, T.W., Hougey, E.H., Jones, L.G. and Mason, J.W., 1971, Renin, norepinephrine and epinephrine responses to graded exercise. *J. Appl. Physiol.*, **31**(2), 178–84.

Kotlowski, A., Krynicki, A., Laba, L., Grimmer, G., Harms, U., Jankowski, A., Mrozinski, W., Grzegorzewski, M., Popinski, J. and Tomaszunas, S., 1986, Morbidity among the seamen of the Polish Merchant Marine and the Merchant Fleet of the G.D.R. in 1974–1976. *Bull. Inst. Marit. Trop. Med.* Gdynia, **37**, 17.

Krasavina, T., Volkova, T. and Gornik, V., 1977, Fundamental condition of the sympatho-adrenal system in patients with late stages of vibration disease. *Klin. Med* (Moscow), **55**, 77.

Krieger, G.R., 1987, Shift work studies provide clues to industrial accidents. *Occupat. Health and Safety*, **56**(1), 20.

LaDou, J., 1982, Health effects of shift work. *Western J. Med.*, **137**(6), 525.

Lambertsen, C.J. (ed.), 1967, *Underwater Physiology*, (Baltimore: Williams & Wilkins Co.).

Lamke, L.O., Lennquist, S., Liljedahl, S.O. and Wedin, B., 1972, The influence of cold stress on catecholamine excretion and oxygen uptake of normal persons. *Scand. J. Clin. Lab. Invest.*, **30**, 57–62.

Lammert, O., 1972, Maximal aerobic power and energy expenditure of Eskimo hunters in Greenland. *J. Appl Physiol.*, **33**, 284.

Larson, P.S., Haag, H.B. and Silvette, H., 1961, *Tobacco* (Baltimore: Williams & Wilkins Co.).

Laurig, W. and Smolka, R., 1980, Development of a portable storage device and an output unit for digital data collection of electrical muscular activity in industrial medicine and work physiology. *Biomedizinsche Technik*, **25**(10), 247.

Lawrie, J.A., 1985, Comparison of past with present, in *Developments in Diving Technology* (Soc. for Underwater Technology) (London: Graham and Trotman).

LeBlanc, J., Côté, J., Dulac, S. and Dubon-Turcot, F., 1978, Effect of age, sex and physical fitness on responses to local cooling. *J. Appl. Physiol.*, **44**, 813.

Lehman, G., 1953, Praktische Arbeits-Physiologie (Stuttgart: Georg Thieme Verlag).

Leithead, C.A. and Lind, A.R., 1964, *Heat Stress and Heat Disorders* (London: Cassell).

Lennquist, S.A., 1972, A comparison between the renal responses to cold and to psychological stressors in human subjects in climatic chamber under standardized conditions. *Swed. J. Def. Med.*, **8**, 70.

Levi, L., 1967, Sympatho-adrenomedullary responses to emotional stimuli: methodologic, physiologic and pathologic considerations, in E. Bajusz (ed.), *An Introduction to Clinical Neuroendocrinology* (New York: S. Karger), 78–105.

Levi, L., 1968, Sympatho-adrenomedullary and related biochemical reactions during experimentally induced emotional stress, in R.P. Michael, (ed.), *Endocrinology and Human Behaviour*, (London: Oxford University Press).

Lewin, R. 1981, Ethiopian stone tools are world's oldest. *Science*, **211**, 806.

Lind, A.R. and Bass, D.E., 1963, Optimal exposure time for development of acclimatization to heat. *Fed. Proc.*, **22**, 704.

Ljungberg, A-S., Enander, A. and Holmér, I., 1979, Evaluation of heat stress during sedentary work. *Scand. J. Work Environ. Health*, **5**(1), 23.

Locati, G., Vicini, A. and Secondi, E., 1974, Correlation between various microclimatic indices. Results obtained from the study of 200 work places. *Med. Lavoro*, **65**(11–12), 451.

Louhevaara, A., Ilmarinen, J. and Ojal, P., Comparison of three field methods for measuring oxygen consumption. *Ergonomics*, **28**(2), 463.

Luczak, H., 1979, Fractioned heart rate variability. Part II. Experiments on superimposition of components of stress. *Ergonomics*, **22**(12), 1315.

Luksza, A.R., Bennett, P. and Earis, J.E., 1985, Bird fancier's lung: hazard of the fishing industry. *Brit. Med. J.*, **291**(6511), 1766.

Lund-Larsen, P.G. and Dahlberg, B.E., 1981, Blodtrykk hos ansatte i spesielt varme og i vanlige avdelinger ved aluminiumsmelteverket i Årdal. *Tidsskrift for Den Norske Laegeforening*, Nr. 33, 1892.

Maasen, A., Meers, A. and Verhaegen, P., 1980, Quantitative and qualitative aspects of sleep in young self-selected four-shift workers. *Int. Arch. Occup. Environ. Health*, **45**(1), 81.

Macleod, J., 1967, Male infertility, in *Advances in Gynecology*, vol. 1 (Baltimore: Williams and Wilkins).

Macleod, J. and Hotchkiss, R.S., 1941, The effect of hyperpyrexia upon spermatozoa counts in men. *Endocrinology*, **28**.

Maehlum, S., Bolling, A., Huser, P.O., Jebens, E. and Tenfjord, O., 1978, *arbeids-fysiologisk undersøkelse av varmebelastningen om bord i M/S Tarn i fart på den Persiske Gulf. 3 S-rapport-serien*, Oslo.

Magnus, P., Bolling, A., Eide, I., Enger, N., Gundersen, N., Guthe, T., Jebens, E., Knutsen, K.E. and Rodahl, K., 1980, Varmestress i ferrolegeringsindustrien. En orienterende undersøkelse ved Fiskaa Verk, Kristiansand. Arbeidsfysiologisk Inst., Oslo.

Mairiaux, P., Nullens, W., Fesler, R., Brasseur, L. and Detry, J.M., 1977, Evaluation des effects d'un vetement refroisissant sur l'adaptation aux efforts prolonges realises a haute temperature par des ouvriers mineurs. *Rev. Inst. Hygiene mines*, **32**, 99.

Makarenko, N.A., 1969, Vegetative nervous system in patients with vibration disease. *Gig. Tr. Prof. Zabol.* **13**, 5.

Malchaire, J.B., 1976, Evaluation of natural wet bulb and wet globe temperature. *Am. Occup. Hyg.*, **19**, 251.

Malm, O.J., 1968, Stress: Fysiologiske og biokjemiske aspekter. *Mensen-dieckbladet*, No. 3, p. 3.

Malmqvist, R., Ekblom, I., Lindström, L., Petersén, I. and Ørtengren, R., 1981, Measurement of localized muscle fatigue in building work. *Ergonomics*, **24**(9), 695.

Martin H. de V. and Callaway, S., 1974, An evaluation of the heat stress of protective face mask. *Ergonomics*, **17**, 221.

Mason, J.W., 1968, A review of psychoendocrine research on the pituitary-adrenal system. *Psychosom. Med.*, **30**, 576.

Matoba, T. and Sakurai, T., 1987, Physiological methods used in Japan for the diagnosis of suspected hand-arm vibration syndrome. *Scand. J. Work Environ. Health*, **13**(4), 334.

Maughan, R.J., 1984, Relationship between muscle strength and muscle cross-sectional area. *Sports Med.*, **1**, 263.

Maxwell, V.B., Crump, J.H. and Thorp, J., 1983, The measurement and risk indicators for coronary heart disease in air traffic control officers: a screening study in a healthy population. *Aviat. Space Environ. Med.*, **54**(3), 246–9.

Melton, C.E., Smith, R.C., McKenzie, J.M., Wicks, S.M. and Saldivar, J.T., 1978, Stress in air traffic personnel: low-density towers and flight service stations. *Aviat. Space Environ. Med.*, **49**(5), 724–8.

Melton, C.E., McKenzie, J.M. and Polis, B.D., 1979, *Physiological Responses in Air Traffic Control Personnel: O'Hare Tower Report FAA-AM-71-2*, January.

METROSONICS, 1981, Metrologger System, db-301/651. Brochure from Metrosonics Inc., Box 18090, Rochester, N.Y.

Millar-Craig, M.W., Bishop, C.N., and Raftery, E.B., 1978, Circadian variation of blood pressure. *The Lancet*, April 15, 795.

Miller, G., 1987, Exposure guidelines for magnetic fields. *Am. Ind. Hyg. Assoc. J.*, **48**(12), 957–68.

Mink, J.W., Blumenschine, R.J. and Adams, D.B., 1981, Ratio of central nervous system to body metabolism in vertebrates: its constancy and functional basis, *Am. J. Physiol.*, **241**, R203.

Minors, D.S. and Waterhouse, J.M., 1985, Introduction of circadian rhythms, in S. Folkard, and T.H. Monk (eds), *Hours of Work* (New York: John Wiley and Sons).

Miyashita, K., Shiomi, S., Itoh, N., Kasamatsu, K. and Iwata, H., 1983, Epidemiological study of vibration syndrome in response to total hand-tool-operating time. *Brit. J. Ind. Med.*, **40**, 92.

Mjøs, L. and Jevnaker, B.H., 1978, Samarbeid i Østerøy-industrien, Arbeidsrapport nr. 22, Industri-økonomisk Institutt, Bergen.

Mjøs, L., Sørlie J.E., and Øvsthus, A., 1982, *Fem små mekaniske verksteder*, Oslo: Universitetsforlaget.

Mohler, S.R., 1983, The human element in air traffic control: aero-medical aspects, problems and prescriptions. *Aviation, Space and Environ. Med.*, **54**(6), 511.

Mohler, S.R., Dille, J.R. and Gibbons, H.L., 1968, The time zone and circadian rhythms in relation to aircraft occupants taking long-distance flights *Am. J. Public Health*, **58**(8), 1404.

Moore, J.W. and Newbower, R.S., 1978, Noncontact tympanic thermometer. *Med. & Biol. Eng. & Comp.*, **16**(5), 580.

Morales, I.V. and Konz, S., 1968, The physiological effect of a water cooled hood in a heat stress environment. *ASHRAE Trans.*, **74**(2), 49.

Morehouse, L.E., 1972, *Laboratory Manual for Physiology of Exercise* (St. Louis: C.V. Mosby Co.).

Müller, E.A. and Franz, H. 1952, Energieverbrauchsmessungen bei beruflicher

Arbeit mit einer verbesserten Respirations-Gasuhr. *Arbeitsphysiol.*, **14**, 499.

Muller, K.E. and Barton, C.N., 1987, A nonlinear version of the Coburn Forster and Cane model of blood carboxyhemoglobin. *Atmosph. Environ.*, **21**(9), 1963.

Mundal, R., Erikssen, J. and Rodahl, K., 1982, Latent ischemic heart disease in sea captains. *Scand. J. Work Environ. Health*, **8**, 178.

Mundal, R., Ramstad, H., Erikssen, J. and Rodahl, K., 1983, Health and Stress of Air Traffic Controllers, Proc. XXXI International Congress of Aviation and Space Medicine, Amsterdam, 202.

Mundal, R., Erikssen, J. and Rodahl, K., 1987, Assessment of physical activity by questionnaire and personal interview with particular reference to fitness and coronary mortality. *Eur. J. Appl. Physiol.*, **56**, 245.

Murphy, T.J., 1979, Fixed vs. rapid rotation shift work. *J. Occup. Med.*, **21**(5), 319.

Mutchler, J.E. and Vecchio, J.L., 1977, Empirical relationships among heat stress indices in 14 hot industries. *Am. Ind. Hyg. Assoc. J.*, **38**(6), 253.

Mutchler, J.E., Malzahn, D.D., Vecchio, J.L. and Soule, R.D., 1976, An improved method for monitoring heat stress levels in the work place. *Am. Ind. Hyg. Assoc. J.*, **37**(3), 151.

Naitoh, P., Englund, C.E. and Ryman, D., 1982, Restorative power of naps in designing continuous work schedules, *J. Human Ergol.*, 11 (Supplement), 259.

Newhouse, M.L., 1966, Dogger Bank itch: survey of trawlermen. *Brit. Med. J.*, **1**, 1142.

Ng, S.C., Chao, T.C. and How, J., 1978, Death by accidental drowning in Singapore, 1973–76. *Singapore Med. J.*, **19**(1), 14.

Nielsen, B., 1969, Thermoregulation in rest and exercise. *Acta Physiol. Scand.* (Suppl. 323).

Nielsen, B., 1980, Exercise and temperature regulation, in Z. Szelényi and M. Székely (eds), Satellite of the 28th International Congress of Physiological Sciences, Péces, 537.

Nilsson, S., (1970), Snurrevadfisket. *Tidsskrift for Den norske Laegeforening*, **90**, 1375.

NIOSH, 1972, *Occupational Exposure to Hot Environments, Criteria for a Recommended Standard.* U.S. Dept. of Health, Education, and Welfare, Nat. Inst. for Occupational Safety and Health, Cincinnati.

Nunneley, S.A. and Flick, C.F., 1981, Heat stress in the A-10 cockpit: flights over desert. *Aviat. Space Environ. Med.*, **52**, 513.

Nunneley, S.A., Webb, P. and Troutman, S. J., 1970, Head cooling during work and heat stress. *Ergonomics*, **13**(4), 527.

Oberoi, K., Dhillon, M.K. and Miglani, S.S., 1983, A study of energy expenditure during manual and machine washing of clothes in India. *Ergonomics*, **26**(4), 375.

O'Donnell, D.R., 1971, The internal clock of the international jet traveller. *Med. J. Australia*, **1**(23), 1227.

Ohashi, N. and Sugihara, Y., 1967, Psychische Anstrengung des Schiffsführers (1). über Psychische Anstrengung bei Abfahrt und Ankunft Grossen Schiffs. *J. Science of Labour*, **43**(8).

Ohashi, N. and Hirota, Y., 1969, *The Mental Strain of a Ship Manoeuver*. XVI Internat. Congress on Occup. Health, Tokyo, Japan, Sept. 22–27.

Oleinikov, V.A., 1975, Physiological aspects of the work of miners on working faces. *Human Physiology*, **4**(5), 677.

Olesen, B., Sliwinska, E., Madsen, T. and Fanger, P.O., 1982, Effect of body posture and activity on the insulation of clothing. Measurement by a movable thermal manikin. *ASHRAE Transactions*, **32**, 791.

Olumide, Y.M., Oleru, G.V. and Enu, C.C., 1983, Cutaneous implications of excessive heat in the work-place. *Contact Dermatitis*, 1983, **9**(5), 360–3.

Opstad, P.K., 1987, Psychological and Physiological Alterations during Strenuous Military Activities. The Significance of Sleep, Abstract. Conf. on Sleep, Lyon.

Östberg, O., 1973, Circadian rhythms of food intake and oral temperature in 'morning' and 'evening' groups of individuals. *Ergonomics*, **16**(2), 203.

Otterland, A., 1970, *Accidents among Merchant Seamen*, Abbotempo, Book 2, 20.

Paffett, J.A.H., 1978, Down to the sea in simulators. *Spectrum*, No. 153, p. 14.

Palca, J.W., Walker, J.M. and Berger, R.J., 1981, Tympanic temperature during rapid eye movement sleep in humans exposed to cold. *Cryobiology*, **18**(1), 92.

Parikh, D.J., Pandya, C.B. and Ramanathan, N.L., 1976, Applicability of the WBGT index of heat stress to work situations in India. *Indian J. Med. Res.*, **64**, 327.

Passmore, R. and Durnin, J.V.G.A., 1955, Human energy expenditure, *Physiol. Rev.*, **35**, 801.

Pasternack, A., 1978, Die Kuhlweste-ein weiterer schritt in der Human isierung des Arbeitslebens. *Dragerheft*, no. 310, 17.

Pavlenko, M.E., 1971, Cardiovascular morbidity among workers in hot shops. *Vrach. Delo*, **6**, 1.

Pepler, R.D., 1963, Performance and well-being in heat, in J.D. Hardy (ed.), *Temperature: Its Measurements and Control in Science and Industry, Vol. 3, Part 3* (New York: Reinhold), 319.

Peterson, J.E., 1970, Postexposure relationship of carbon monoxide in blood and expired air. *Arch. Environ. Health*, **21**, 172.

Peterson, J.E. and Stewart, R.D., 1970, Absorption and elimination of carbon monoxide by inactive young men, *Arch. Environ. Health*, **21**, 165.

Peterson, J.E. and Stewart, R.D., 1975, Predicting the carboxyhaemoglobin level resulting from carbon monoxide exposures. *J. Appl. Physiol.*, **39**(4), 633.

Plunkett, J.M. and Carter, R.P., 1974, Practical problems in the use of WBGT for heat stress evaluation. *Am. Ind. Hyg. Assoc. J.*, **35**(5), 287.

Polish Ergonomics Society, 1979, *Ergonomia*, **2**(1).

Praml, G.J., Mauermayer, R. and Hartmann, L., 1986, Personenbezogene Erfassung des zeitlichen Verlaufes der Staubbelastung mit einen Tyndallometer und digitaler Speicherung, STAUB. *Reinhaltung der Luft*, **46**(3), 140.

Precht, H., Christophersen, J., Hensel, H. and Larcker, W., 1973, *Temperature and Life* (Berlin: Springer-Verlag).

Preston, F.S. and Bateman, S.C., 1970, Effect of time zone changes on the sleep patterns of BOAC B.707 crews on world-wide schedules. *Aerospace Med.*, **41**(12), 1409.

Pruett, E.D.R., 1971, *Fat and Carbohydrate Metabolism in Exercise and Recovery, and its Dependence upon Work Load Severity*. Inst. of Work Physiology, Oslo.

Pulket, C., Henschel, A., Burg, W.R. and Saltzman, B.E., 1980, A comparison of heat stress indices in a hot-humid environment. *Am. Ind. Hyg. Assoc. J.*, **41**(6), 442.

Radigan, L.R. and Robinson, S., 1949, Effects of environmental heat stress and exercise on renal blood flow and filtration rate. *J. Physiol.* **4**, 185.

Ramanathan, L.N., 1964, A new weighing system for mean surface temperature of the human body. *J. Appl. Physiol.*, **19**(3), 531.

Ramsey, J.D., 1975, Occupational vibration, in C. Zenz (ed.), *Occupational Medicine: Principles and Practical Applications* (Chicago: Year Book Medical Publishers), 553–562.

Ramsey, J.D. and Chai, C.P., 1983, Inherent variability in heat-stress decision rules. *Ergonomics*, **26**(5), 495.

Rapp, R. and Aubertin, G., 1985, Thermal load prediction – calculation and measurement of mean radiation temperature in workshops. *Cahiers de notes documentaires – Securité et hygiene du travail*, **121**, 429.

Rea, J.N., Tyrer, P.J., Kasap, H.S. and Beresford, S.A.A., 1973, Expired air carbon monoxide, smoking, and other variables. A community study, *Brit. J. Prev. Soc. Med.*, **27**, 114.

Rees, P.J., Chilvers, C. and Clark, T.J.H., Evaluation of methods used to estimate inhaled dose of carbon monoxide. *Thorax*, **35**, 47.

Reinberg, A., Vieux, N., Ghata, J., Chaumont, A.J. and Laporte, A., 1978, Circadian rhythm amplitude and individual ability to adjust to shift work. *Ergonomics*, **21**(10), 763.

Rembowski, J., 1981, Emotional relations in two-children seamen families. *Bull. Inst. Marit. Trop. Med. Gdynia*, **32**, 191.

Resmond, C.K., Emes, J.J., Mazumdar, S., Magee, P.C. and Kamon, E., 1978, Morality of steelworkers employed in hot jobs. *J. Environ. Pathology & Toxicology*, **2**(5), 75–96.

Riggs, C.E., Johnson, D.J., Konopka, B.J. and Kilgour, R.D., 1981, Exercise heart rate response to facial cooling, *Eur. J. Appl. Physiol.*, **47**(4), 323.

Riggs, C.E., Johnson, D.J., Kilgour, R.D., and Konopka, B.J., 1983, Metabolic effects of facial cooling in exercise. *Aviat. Space Environ. Med.*, **54**(1), 22.

Robinson, S., 1963, Circulatory adjustments of man in hot environments, in J.D. Hardy (ed.), *Temperature: Its Measurement and Control in Science and Industry, Vol. 3, Part 3*, (New York: Reinhold Book Corp.), p. 287.

Rodahl, A., O'Brien, M. and Firth, R.G.R., 1976, Diurnal variation in performance of competitive swimmers, *J. Sports Med. and Physical Fitness*, **16**, 72.

Rodahl, K., 1953, *North. The Nature and Drama of the Polar World*, (New York: Harper & Bros.).

Rodahl, K., 1963, Nutritional Requirements in the Polar Regions. U.N. Symposium on Health in the Polar Regions, Geneva, 1962, *WHO Public Health Paper*, **18**, 97.

Rodahl, K., 1980, Arbeidsstress til Sjös, System for Sikkert Skip. NTNF Rapport 80/2 ISMN82-7224-043-2, Norges Teknisk-Naturviten-skapelige Forskningsråd, Oslo 1980.

Rodahl, K., 1981, Heat stress: Norwegian experience in health protection in primary aluminium production, in *IPAI, vol. 2*, ed. J.P. Hughes (London).

Rodahl, K. and Guthe, T., 1988, Physical limitations of human performance in hot environments with particular reference to work in heat-exposed industry, in J.B. Mekjavic, E.W. Banister and J.B. Morrison (eds), *Environmental Ergonomics* (London: Taylor and Francis).

Rodahl, K. and Huser, P.O., 1978, *Arbeidsfysiologisk undersøkelse på M/S Bergensfjord, med saerlig henblikk på brooffiserene.* 3 S-rapportserien, Oslo.

Rodahl, K. and Issekutz jr., B., 1962, Physical performance capacity of the older individual, in K. Rodahl and S.M. Horvath (eds), *Muscle as a Tissue* (New York: McGraw-Hill Book Co.).

Rodahl, K. and Vartdal, F., 1977, *En arbeidsfysiologisk analyse av belastningen om bord på taubåten 'Tender Pull' under ankerhåndtering på oljefeltet i Nordsjøen.* 3 S-rapportserien, Oslo.

Rodahl, K. and Vokac, Z., 1977a, Work stress in Norwegian trawler fishermen. *Ergonomics*, **20**(6), 633.

Rodahl, K. and Vokac, Z., 1977b, The work physiology of fishing. *Nordic Council Arct. Med. Res. Rep.* No. **18**, 22.

Rodahl, K., Bolling, A. and Tenfjord, O., 1979, *Arbeidsfysiologisk undersøkelse om bord på M/T Texaco Bergen.* 3 S-rapportserien, Oslo.

Rodahl, K., Miller, H.I. and Issekutz, B., jr., 1964, Plasma free fatty acids in exercise, *J. Appl. Physiol.*, **19**, 489.

Rodahl, K., Tenfjord, O. and Vartdal, F., 1977b, med teknisk assistanse av P.O. Huser og. J. Rodahl, En orienterende arbeidsfysiologisk undersøkelse av arbeidsplassene om bord på containerskipet M/S Toyama. 3 S-rapportserien, Oslo.

Rodahl, K., Bolling, A., Olsen, B. and Jebens, E., 1980a, *En arbeidsfysiologisk undersøkelse på hurtigruteskipet Harald Jarl.* 3 S-rapport serien, Oslo.

Rodahl, K., Giere, F.A., Staff, P.H. and Wedin, B., 1974a, A physiological comparison of the protective value of nylon and wool in a cold environment, in A. Borg and J.H. Veghte (eds), *AGARD Report*, no. 620.

Rodahl, K., Horvath, S.M., Birkhead, N.C. and Issekutz, jr., B., 1962, Effects of dietary protein on physical work capacity during severe cold stress. *J. Appl. Physiol.*, **17**, 763.

Rodahl, K., Mundal, R., Huser, P.O. and Bjørgan, P., 1985, Ledere for små og mellomstore bedrifter. En arbeidsfyiologisk undersøkelse. Unpublished.

Rodahl, K., Tenfjord, O., Brochman, J. and Jebens, E., 1980b med teknisk assistanse av B. Olsen og J. Rodahl, *En arbeidsfysiologisk undersøkelse på kombinasjonsskipet S/S Docecanyon.* 3 S-rapportserien, Oslo.

Rodahl, K., Vokac, Z., Fugelli, P., Vaage, O. and Maehlum, S., 1974, Circulatory strain, estimated energy output and catecholamine excretion in Norwegian coastal fishermen. *Ergonomics*, **17**(5), 585.

Rode, A. and Shephard, R.J., 1971, Cardiorespiratory fitness of an Arctic community. *J. Appl. Physiol.*, **31**, 519.

Rodgers, S.H. and Eggleton, E.M., (eds.), 1983, *Ergonomic Design for People at Work, Vol. 1.* (New York: Van Nostrand Reinhold Co.).

Rodin, I., Abeytunga, P.K. and Cohen, D., 1984. The occupational health physician and the microcomputer. *J. Soc. Occup. Med.*, **34**, 27.

Rognum, T.O. and Erikssen, J., 1980, Blood pressure elevation caused by stress at work. *Tidsskrift for Den norske laegeforening*, **3**, 168.

Rohmert, W., 1968, Die Beziehung Zwischen Kraft und Ausdauer bei Statischer Muskelarbeit, Schriftewreike Arbeitsmedizin, Sozialmedizin, Arbeitshygiene, band 22, (Stuttgart: A.W. Gentner Verlag), 118.

Roland, P.E., 1987, Changes in brain blood flow and oxidative metabolism during mental activity, *News in Physiological Sciences*, **2**, 120–4.

Rönningen, H., 1976, unpublished results.

Roscoe, A.H., 1978, Stress and workload in pilots, *Aviat. Space Environ. Med.*, **49**(4), 630.

Ross, M.H., 1972, The Appalachian coal miner: his way of living, working, and relating to others. *Annal New York Academy of Sciences*, **200**, 184.

Ross, M.H., 1977, Some aspects of health hazards to seafaring men. *Central African J. Med.*, **23**(12), 293.

Rowell, L.B., 1974, Human cardiovascular adjustment to exercise and thermal stress. *Physiol. Rev.*, **54**, 75.

Rutenfranz, J. and Colquhoun, W.P. (eds), 1978, Shiftwork: theoretical issues and practical problems. *Ergonomics*, **21**(10), 737.

Rutenfranz, J. and Colquhoun, W.P., 1979, Circadian rhythms in human performance *Scand. J. Work. Environ. Health*, **5**, 167.

Rutenfranz, J., Wahle, A. and Klimmer, F., 1981, *Ein Hygieniker entwickelt die Arbeitsphysiologie-Geheimrat Rubner und Seine Mitarbeiter-Arbeitsmedizinisches Kolloquium* (Stuttgaart: Gentner Verlag).

Rutenfranz, J., Colquhoun, W.P., Knauth, P. and Ghata, J.N., 1977, Biomedical and psychosocial aspects of shift work. A review. *Scand. J. Work Environ. & Health*, **3**(4), 165.

Saito, K., 1987, Prevention of the hand-arm syndrome. *Scand. J. Work. Environ. Health*, **13**(4), 301.

Saltin, B., 1964, Aerobic work capacity and circulation at exercise in man. *Acta Physiol. Scand.*, **62**, (Suppl. 230).

Saltin, B. and Åstrand, P.-O., 1967, Maximal oxygen uptake in athletes, *J. Appl. Physiol.*, **23**, 353.

Saltin, B., Blomqvist, B., Mitchell, J.H., Johnson, jr., R.L., Wildenthal, K. and Chapman, C.B., 1968, Response to submaximal and maximal exercise after bed rest and training. *Circulation*, **38**, (Suppl. 7).

Sandrackaja, S.E., 1966, Investigations on the crew's working conditions on fishing trawlers and freezer trawlers. *Bull. Inst. Mar. Trop. Med. Gdansk*, **17**, 231 (in Russian).

Schilling, R.S.F., 1971, Hazards of deep-sea fishing. *Brit. J. Industr. Med.*, **28**(1), 27.

Schneider, R.A., 1981, Fully automatic portable recording of indirect blood pressures in subjects awake and asleep, at home and at work, *Biotelemetry & Patient Monitoring*, **8**(1–2), 81.

Scholander, P.F., Walters, V., Hook, R. and Irving, L., 1950, Body insulation of some Arctic and tropical mammals and birds. *Biol. Bull.*, **99**, 225.

Shapiro, Y., Pandolf, K.B., Avellini, B.A., Pimental, N.A. and Goldman, R.F., 1981, Heat balance and transfer in men and women exercising in hot-dry and hot-wet conditions. *Ergonomics*, **24**, 375.

Shephard, R.J., 1984, Tests of maximum oxygen intake. A critical review. *Sports Med.*, **1**, 99.

Shvartz, E., 1970, Effect of a cooling hood on physiological responses to work in a hot environment. *J. Appl. Physiol.*, **29**(11).

Shvartz, E., 1976, Effect of neck versus chest cooling on responses to work in heat. *J. Appl. Physiol.*, **40**(5), 668.

Shvartz, E., Aldjem, M., Ben-Mordechai, J. and Shapiro, Y., 1974, Objective approach to a design of a whole-body, water-cooled suit. *Aerospace Med.*, **45**(7), 711.

Siconolfi, S.F., Cullinane, E.M., Carleton, R.A. and Thompson, P.D., 1982, Assessing VO_2 in epidemiologic studies; modification of the Åstrand-Rhyming test *Med. Sci. Sports Exc.*, **14**(5), 335.

Simonsen, E. and Lind, A.R., Fatigue in static work, in E. Simonsen (ed.), *Physiology of Work Capacity and Fatigue*, (Springfield, IL: Charles C. Thomas), p. 241.

Singer, R. and Rutenfranz, J., 1971 Attitudes of air traffic controllers at Frankfurt airport towards work and the working environment. *Ergonomics*, **14**, 633–9.

Siple, P.A. and Passel, C.F., 1945, Dry atmospheric cooling in sub-freezing temperatures. *Proc. Am. Philos. Soc.*, **89**, 177.

Snellen, J.W., 1966, Mean body temperature and the control of thermal sweating. *Acta Physiol. Pharmacol. Neerl.*, **14**, 99.

Sørlie, J.E., 1982, Papirløst lederskap, Rapport nr. 21. Industriøkonomisk Institutt, Bergen.

Spitzer, W.O., Hill, G.B., Chambers, L.W., Helliwell, B.E. and Murphy, H.B., 1975, The occupation of fishing as a risk factor in cancer of the lip. *The N. Eng. J. Med.*, **293**(9), 419.

Staff, P.A. and Nilsson, S., 1971, Vaeske og Sukkertilførsel under Langvarig intens Fysisk Aktivitet, *Tidsskrift for Den Norske Laegeforening*, **16**, 1235.

Stewart, R.D., Peterson, J.E., Baretta, E.D., Bachand, R.T., Hosko, M.J. and Herrmann, A.A., 1970, Experimental human exposure to carbon monoxide, *Arch. Environ. Health*, **21**, 154.

Sudo, A., 1980, Urinary excretion of adrenaline, noradrenaline, 17-OHCS, 5-hydroxytryptamine and certain electrolytes in 24-hour shift workers taking a 4-hour night nap. *Industrial Health*, **18**(3), 117.

Sundin, I., 1956, The influence of body posture on the urinary excretion of adrenalin and noradrenalin, *Acta Med. Scand.*, **154**, (Suppl. 313).

Sutton, J.R., Jones, N.L., Griffith, L. and Pugh, C.E., 1983, Exercise at altitude. *Ann. Rev. Physiol.*, **45**, 427.

Szczepanski, C. and Zaborski, L., The usefulness of various types of individual hearing protectors for sailors. *Biol. Inst. Med. Morsk. Gdansk*, **32**(3–4), 213.

Tabor, M., 1982, Shift work: worth the risks? *Occup. Health & Safety*, **51**(4), 42.

Tarquini, B., Cecchettin, M. and Cariddi, A., 1986, Serum gastrin and pepsinogen in shift workers. *Int. Arch. Occup. Environ. Health*, **58**, 99.

Tasto, D.L., 1982, Health and safety consequences of shift work in the food processing industry. *Ergonomics*, **25**(2), 133.

Taylor, N.B.G., Kuehn, L.A. and Howat, M.R., 1969, A direct-reading mercury thermometer for the wet bulb globe temperature index. *Can. J. Physiol. Pharmacol.*, **47**, 277.

Taylor, P.J., 1973, The effects of shift work on workers' health, *Ind. Med. Surg.*, **42**(8), 13.

Tenfjord, O.W., Borchgrevink, Chr. F., Vaa, T., Vellar, O.D. and Aarø, L.E., 1983, Risikofaktorer for koronaer hjertesykdom hos sjaemenn. *Tidsskrift for Den norske Laegeforening*, nr. 34–35–36, 2312.

Thomas, K., 1932, Max Rubner, *Klin. Wschr.* 11, 926–7.

Thomas, N.T., 1974, Physiological and psychological effects of industrial protective clothing. *Ergonomics*, **17**(4), 565.

Tikuisis, P., Madill, H.D., Gill, B.J., Lewis, W.F., Cox, K.M. and Kane, D.M., 1987, A critical analysis of the use of the CFK equation in predicting COHb formation. *Am. Ind. Hyg. Assoc. J.*, **48**(3), 208.

Tilley, A.J., Wilkinson, R.T., Warren, P.S.G., Watson, B. and Drud, M., 1982, The sleep and performance of shift workers. *Human Factors*, **24**(6), 629.

Tochikubo, O., Takashi, O., Miyazaki, N. and Kaneko, Y., 1986, A new portable device for 24-hour recording of ambulatory intra-arterial blood pressure and heart rate in hypertensive patients. *Japanese Heart J.*, **27**(5), 661.

Tola, S., Riihimäki, H., Videman, T., Viikari-Juntura, E. and Hänninen, K., 1988, Neck and shoulder symptoms among men in machine operating, dynamic physical work and sedentary work. *Scand. J. Work Environ. Health*, **14**, 299.

Tolonen, M. (ed.), 1979, *Scand. J. Work. Environ. & Health*, (Suppl. 2).

Tomaszewski, R., Chmielewski, J. and Bielawska-Krasnowiecka, G., 1976, Risk factors for coronary heart disease among seamen, *Bull. Inst. Marit. Trop. Med. Gdynia*, **27**.

Tomaszewski, R., Gandurski, P., Bielawska-Krasnowiecka, G., Hac, E. and Chmielewski, J., 1981, Appraisal of the professional fitness of seamen with cardiovascular diseases hospitalized in the years 1978–1980. *Bull. Inst. Marit. Trop. Med. Gdynia*, **32**, 183.

Uchino, K., Kawashima, Y., Ohiro, M., Watanabe, F., Kimotuki, K., Watanabe, A. and Hattori, M., 1986, Studies on tympanic membrane temperature and thermal protection of survival suits in ice-water. *Internat. J. Biometeorology*, **30**(4), 367.

Vaage, O., 1974, Fluorometric determination of epinephrine and norepinephrine in 1 ml urine introducing dithiotreitol and boric acid as stability and sensitivity improving agents of the trihydroxyindole method. *Biochem. Med.*, **9**, 41.

Vaida, I. and Pafnote, M., 1979, Physical work load in modern mining. *Rev. Morphol-Embryol. Physiol., Physiologie*, **16**(2), 133.

Valentine, J.W., 1978, The evolution of multicellular plants and animals. *Sc. Am.*, **239**(3), 67.

Vangaard, L., 1975, Physiological reactions to wet-cold, *Aviation Space and Environ. Med.*, **46**, 33.

Vangaard, L., 1982, Perifer Varmebalanse, in I. Holmér and J. Sundell (eds), *Arbete i Kallt Klimat*, Stockholm. Arbete och Helsa, Arbetarskyddsverket, Vol. 1.

Vangaard, L. and Nielsen, S., 1977, Arbejdsmiljöet i Dansk Fiskeri. *Ugeskrift for Laeger*, **139**(7), 413.

Veit, R.W., 1971, A portable tape recorder for noise analysis, *Amer. Industr. Hyg. Assoc. J.*, Sept., 578.

Velle, W., 1987, Sex differences in sensory functions. Perspectives in Biology and Medicine, **30**(4), 490–522.

Vogt, J.J., Meyer, J.P., Candas, V., Libert, J.P. and Sagot, J.C., 1983, Pumping effects of thermal insulation of clothing worn by human subjects. *Ergonomics*, **26**(10), 963.

Vokac, Z. and Rodahl, K., 1974, *A Study of Continuous Night Work at the Norwegian Steel Mill at Mo i Rana*, Nordic Council for Arctic Medical Research Report, No. 10.

Vokac, Z. and Rodahl, K., 1975, Field study of rotating and continuous night shifts in a steel mill. Proc. Third International Symposium on Night- and Shiftwork in Dortmund, Oct. 28–31, 1974, in P. Colquhoun, S. Folkard, P. Knauth and R. Rutenfranz (eds), *Experimental Studies in Shiftwork* (Opladen: Westdeutscher Verlag) 168.

Vokac, Z. and Rodahl, K., 1976, *Maximal aerobic power and circulatory strain in Eskimo hunters in Greenland. Nordic Council Arct. Med. Res. Report.* No. 16, 16.

Vokac, Z., Bell, H., Bautz-Holter, E. and Rodahl, K., 1975, Oxygen uptake/heart rate relationship in leg and arm exercise, sitting and standing, *J. Appl. Physiol.*, **39**, 54.

Volle, M., Brisson, G.R., Pérusse, M., Tanaka, M. and Doyan, J., 1979, Compressed work-week: psycho-physiological and physiological repercussions. *Ergonomics*, **22**(9), 1001.

Wahren, J., Jequier, E., Acheson, K. and Schutz, Y., 1983, Influence of cigarette smoking on body oxygen consumption. *Clin. Phys.* **3**(1), 91.

Wald, N.J., Idle, M., Boreham, J. and Bailey, A., 1981, Carbon monoxide in breath in relation to smoking and carboxyhaemoglobin levels. *Thorax*, **36**, 366.

Wallace, J., Sweetnam, P.M., Warner, C.G., Graham, P.A. and Cochrane, A.L., 1971, An epidemiological study of lens opacities among steel workers. *Brit. J. Indust. Med.*, **28**, 265.

Waersted, M., Björklund, R. and Westgaard, R.H., 1987, Generation of muscle tension related to a demand of continuing attention, in B. Knave and P.-G. Widebäck (eds), *Work with Display Units*, **86** (North Holland: Elsevier Science Publishers B.V.).

Waersted, M., Björklund, R.A., and Westgaard, R.H., 1988, Muscle Tension induced by Two VDU-Based Tasks Demanding Minimal Motor Activity. Poster presented at IVth Int. Conf. Psychophysiology, Prague, Sept. 12–17.

Washburn, S.L., 1978, The evolution of man. *Sc. Am.*, **239**(3), 146.

Waskiewicz, J. and Banaszkiewicz, T., 1981a, The evaluation of the state of the circulatory system in the officer personnel of the Polish Ocean Lines. I. A general characterization of the investigated population. *Bull. Inst. Marit. Trop. Med.* Gdynia, **32**, 153.

Waskiewicz, J., and T. Banaszkiewicz, 1981b., The evaluation of the state of the circulatory system in the officer personnel of the Polish Icean Lines. II. Changes in ECG rest recording and in phonocardiographic recording. *Bull. Inst. Marit. Trop. Med.* Gdynia, **32**, 159.

Wattkins, S.A., 1986, Leptospirosis as an occupational disease. *Brit. J. Ind. Med.*, **43**(11), 721.

Westgaard, R.H., Waersted, M., Jansen, T. and Korsund, K., 1984, Belastninger og belastningslidelser hos produksjonsarbeidere ved Helly-Hansen A/S, Arbeidsfysiologisk Inst., Oslo.

Westgaard, R.H., Waersted, M., Jansen, T. and Aarås, A., 1986, Muscle load and illness associated with constrained body postures, in N. Corlett, J. Wilson and I. Manenica (eds), *The Ergonomics of Working Postures, Models, Methods and Cases* (London: Taylor & Francis).

Wever, R.A., 1985, Man in temporal isolation, in S. Folkard and T.H. Munk (eds), *Hours of Work* (New York: John Wiley and Sons).

White, P.C. jr., 1977, An evaluation of health standards and practices in the United States Merchant Marine, *Military Medicine*, **142**(9), 706.

Williams, C.G., Wyndham, C.H. and Morrison, J.F., 1967, Rate of loss of acclimatization in summer and winter. *J. Appl. Physiol.*, **22**, 21.

Wilson, G.D. and Sklenka, M.P., 1983, A system of measuring energy cost during highly dynamic activities, *J. Sports Med.*, **23** 155.

Winget, C.M., DeRoshia, C.W., Markley, C.L. and Holley, D.C., 1984, A review of human physiological and performance changes associated with desynchronosis of biological rhythms. *Aviat. Space, & Environ. Med.*, **55**(12), 1085.

Wøhni, B., Westgaard, R.H., Bjørklund, R.A., Jarmark-Robertsson, B. and Rodahl, K., (unpublished), En sammenligning mellom skipsførerens reaksjoner under manøvrering på brosimulator med den samme skips førers reaksjon under manøvrering av samme skip til samme havn i virkeligheten.

Wolf, S. and Wolf, H., 1943, *Human Gastric Function* (New York: Oxford University Press).

Wolff, H.S., 1958, The integrating motor pneumotachograph: a new instrument for the measurement of the energy expenditure by indirect calorimetry. *Quart. J. Exper. Physiol.*, **43**, 270.

Wolthuis, R., Hull, D. McAfoose, D. and Fischer, J., 1981, Portable blood pressure measurements. Performance of Korotko sound analysis techniques. *Hypertension*, **3**, 596.

Wurster, R.D., 1968, Silicon rubber earmould for tympanic membrane temperature, *Pflügers Archiv*, **302**(3), 275.

Wyndham, C.H., Strydom, N.B., Benade, A.J.S. and Rensburg, A.J., 1973, Limiting rates of work for acclimatization at high wet bulb temperatures. *J. Appl. Physiol.*, **35**, 454.

Wyon, D., 1982, Kyla och prestation, in I. Holmér and J. Sundell (eds), *Arbete i kallt klimat* (Stockholm: *Arbete och Hälsa, Arbetarskyddsverket*), Vol. 1.

Wyon, D.P., Andersen, I. and Lundqvist, G.R., 1979, The effects of moderate heat stress on mental performance. *Scand. J. Work Environ. Health*, **5**, 352.

Yaglou, C.P. and Minard, D., 1957, Control of heat casualties at military training centres. *Arch. Ind. Health*, **16**, 302.

Zal, H.A., 1984, Recommended program for employees exposed to extremes of heat. *Occup. Health Nursing*, **32**(6), 293.

Zijlstra, W.G., Buursma, A. and Zwart, A., 1988, Performance of an Automated six-wavelength photometer (Radiometer OSM3) for routine measurement of hemoglobin derivations. *Clin. Chem.*, **34**(1), 149.

Zugaj, A. and Chmielewski, J., 1974, Preliminary investigations on the behaviour of oscillometric indexes in the lower extremities of crew members of a fishing vessel during their voyages. *Bull. Inst. Maritime and Tropical Med. Gdynia.*, **25**(1), 97.

Index